PRENTICE-HALL INTERNATIONAL, INC., *London*
PRENTICE-HALL OF AUSTRALIA, PTY. LTD., *Sydney*
PRENTICE-HALL OF CANADA, LTD., *Toronto*
PRENTICE-HALL OF INDIA PRIVATE LIMITED, *New Delhi*
PRENTICE-HALL OF JAPAN, INC., *Tokyo*

 FOUNDATIONS OF MODERN BIOCHEMISTRY SERIES

Lowell Hager and Finn Wold, editors

ORGANIC CHEMISTRY OF BIOLOGICAL COMPOUNDS*
Robert Barker

INTERMEDIARY METABOLISM AND ITS REGULATION
Joseph Larner

PHYSICAL BIOCHEMISTRY
Kensal Edward Van Holde

MACROMOLECULES: STRUCTURE AND FUNCTION
Finn Wold

SPECIAL TOPICS:

BIOCHEMICAL ENDOCRINOLOGY OF THE VERTEBRATES
Earl Frieden and Harry Lipner

* Published jointly in Prentice-Hall's *Foundations of Modern Organic Chemistry Series.*

PHYSICAL
BIOCHEMISTRY

KENSAL EDWARD VAN HOLDE

Professor of Biophysics
Oregon State University

Prentice-Hall, Inc., Englewood Cliffs, New Jersey

To BARBARA, PAT, MARY, STEVE, and DAVE

© 1971 by Prentice-Hall, Inc., Englewood Cliffs, New Jersey

Current printing (last digit)

10 9 8 7 6 5 4 3 2

Library of Congress
Catalog Card No. 73-158777

13-665885-7 (C)

13-665877-6 (P)

Printed in the United States of America

FOREWORD

Biochemistry has been and still is the major meeting place for biological and physical sciences, and most introductory courses in biochemistry have a rather unique and heterogeneous population of advanced students from all branches of the natural sciences. Such courses, therefore, have equally unique and hetero- geneous requirements for background material and textbooks. The content of a first-year course in biochemistry based on two years of chemistry and biology would probably not be too difficult to define, but to write a single text which contains the needed material for all students becomes a much more controversial issue. As a solution to this dilemma, presenting the material in several packages offers some interesting possibilities. Without in any way compromising the basic content, such a multivolume text should above all allow for a great deal of flexi- bility: flexibility for the student to supplement previous experience with only the parts which represent new and unique features; flexibility for the instructor to offer a course covering an area of modern biochemistry with something less than the comprehensive text; and the flexibility to keep a text current by rewriting any outdated part without having to reject the whole text.

These thoughts were fundamental in formulating the Foundations of Modern Biochemistry series. So was the philosophy that a major purpose of a textbook is perhaps not so much to be encyclopedic as to select for the student the im- portant principles and to illustrate and explain these in some depth. With this basic plan in mind, the next step was to divide the whole into logical subdivisions or parts. The science of biochemistry seeks the answer to three basic questions:

1. What is the nature of the molecules and structures found in living cells?
2. What is the biological function of these molecules and structures?
3. How are they synthesized (and broken down) in the cell?

These questions were adopted as the basis for the division of this text. The first question, related to the qualitative and quantitative characterization of the biochemical world and to the methods available for structural analysis, is covered in two books: *Organic Chemistry of Biological Compounds* and *Physical Biochemistry*. The second question, concerning the elucidation of the biological function of these molecules, is discussed in *Macromolecules: Structure and Function*. The third question, covering all aspects of intermediate metabolism and metabolic regulation, is considered in *Intermediary Metabolism and Its Regulation*. As the work on the individual books progressed, it became apparent that this subdivision had one unexpected advantage; namely that the physical presentation of material as diversified as the mathematical derivations in physical chemistry, the many structures of organic chemistry, and the long and complicated road maps of intermediate metabolism can be handled much more rationally in the individual format of separate books than in a single volume.

Thus, the Foundations of Modern Biochemistry series came into being consisting of four *individual* books. The books can hopefully be used either separately or in any combination to meet the requirements of the individual student according to his particular background and goals. The absence of a numbered sequence represents a deliberate effort to emphasize that the books can be used in any order. The integration of the four parts into the total program is done by extensive cross-references between the individual books. Trying to retain the individuality and utility of the separate books and at the same time aiming for an integrated program required some compromises in the selection and distribution of the material to be covered. Quite expectedly, the main price paid for this dual purpose has turned out to be some duplication, which hopefully is not extensive enough to become a serious flaw.

In arriving at the definition of the Foundations of Modern Biochemistry content and at the philosophy represented by this set of books, the thinking of the "editorial board" (the four authors and Dr. Lowell P. Hager) underwent an extensive evolution. The evolutionary pressures were graciously, patiently, and sometimes even enthusiastically provided by colleagues, by the publishers, and most importantly by students, too many to mention individually. To all of these good people we express our sincere thanks.

FINN WOLD

CONTENTS

INTRODUCTION

Cell and tissue, shell and bone, leaf and flower, are so many portions of matter, and it is in obedience to the laws of physics that their particles have been moved, moulded, and conformed.

D'ARCY THOMPSON, 1917†

These words tell what molecular biology is all about. By proceeding from the assumption that the fundamental physical and chemical mechanisms of life are knowable, it has become possible to understand, at least in part, how inheritance works, how cells make proteins, how proteins regulate metabolism, how muscles contract, and much more. Because the fusion of physics, chemistry, and biology we call molecular biology has been so fruitful, it has come to dominate much of biochemistry.

The molecular interpretation of biological events shows every indication of becoming still more detailed and powerful; therefore the modern student of biochemistry *must* become acquainted with physico-chemical theories and methods. These are the subject of this volume. The treatment given here is only a sketch, a preliminary look intended to acquaint the student with a number of ideas and explore a few. It dwells, perhaps too much, on the physico-chemical tools the modern biochemist uses. But an understanding of the techniques is fundamental to their intelligent use.

This book presupposes some introduction to physical chemistry. Precisely, I assume that the student has had at least a one-semester course, in which the fundamentals of thermodynamics have been introduced. Having taught such

† *On Growth and Form*, Cambridge University Press, Cambridge, 1917, p. 7.

1

courses, I am not deluded about the depth of understanding obtained. For this reason, thermodynamics is reviewed in the first chapter.

A word of advice to those who wish to make a career of biochemical research : Go back and learn deeply the fundamentals of mathematics, physics, and chemistry. Then you will be able to show us precisely how the particles of life are "moved, moulded, and conformed."

ONE | THERMODYNAMICS AND BIOCHEMISTRY

Thermodynamics, with its emphasis on heat engines and abstract energy concepts, has often seemed irrelevant to biochemists. Indeed, a conventional introduction to the subject is almost certain to convince the student that much of thermodynamics is sheer sophistry and unrelated to the real business of biochemistry, which is discovering how molecules make organisms work.

But an understanding of some of the ideas of thermodynamics *is* important to biochemistry. In the first place, the very abstractness of the science gives it power in dealing with poorly defined systems. For example, one can use the temperature dependence of the equilibrium constant for a protein denaturation reaction to measure the enthalpy change without knowing what the protein molecule looks like or even its exact composition. And the magnitude and the sign of that change tell us something more about protein molecules. Again, the modern biochemist continually uses techniques that depend on thermodynamic principles. He may measure the molecular weight of a macromolecule or study its self-association by osmotic pressure measurements. All he observes is a pressure difference, but he *knows* that this difference can be quantitatively interpreted to yield an average molecular weight. To use these physical techniques intelligently, he must understand something of their bases; this is what a good deal of this book is about.

In this first chapter we shall briefly review some of the ideas of thermodynamics that are of importance to biochemistry and molecular biology.

3

While most readers will have taken an undergraduate course in physical chemistry, it has been my experience that this usually leads, insofar as thermodynamics is concerned, to a fairly clear understanding of the first law and some confusion about the second. Since the aim of this section is the use of thermodynamics rather than contemplation of its abstract beauty, we shall emphasize some molecular interpretations of thermodynamic principles. But it should never be forgotten that thermodynamics does not depend, for its rigor, on explicit details of molecular behavior. It is, however, sometimes easier to visualize thermodynamics in this way.

1.1 HEAT, WORK, AND ENERGY

First Law of Thermodynamics

The intention of biochemistry is ultimately to describe certain macroscopic systems, involving multitudes of molecules, in terms of individual molecular properties. The fact is, however, that such systems are so complex that a complete description is beyond the capabilities of present-day physical chemistry. On the other hand, a whole field of study of energy relationships in macroscopic systems has been developed that makes no appeal whatsoever to molecular explanations. This discipline, thermodynamics, allows very powerful and exact conclusions to be drawn about such systems. The laws of thermodynamics are quite exact for systems containing many particles, and this gives a clue as to their origin. They are essentially statistical laws.

The situation is in a sense like that confronting an insurance company; the behavior of the individuals comprising its list of insurees is complex, and an attempt to trace out all of their interactions and predict the fate of any one would be a staggering task. But if the number of individuals is very large, the company can rely with great confidence on statistical laws, which say that so many will perish or become ill in any given period. Likewise, the physical scientist can draw from his experience with macroscopic bodies (large populations of molecules) laws that work very well indeed, even though they leave obscure the mechanism whereby the phenomena are produced. Just as the insurance company with only 10 patrons is in a precarious position (it would not be too unlikely for all of them to die next month), so is the chemist who attempts to apply thermodynamics to systems of a few molecules. But 1 mole is 6×10^{23} molecules, a large number indeed. However, a bacterial cell may contain only a small number of some kinds of molecules; this means that some care must be taken when one attempts to apply thermodynamic ideas to systems of this kind. [But see T. L. Hill (1963).]

For review, let us define a few fundamental quantities.

System: A part of the universe chosen for study. It will have spatial boundaries but may be *open* or *closed* with respect to the transfer of matter. Similarly, it may or may not be thermally insulated from its surroundings. If it is, it is said to be an *adiabatic* system.

State of the system: The thermodynamic state of a system is clearly definable only for systems at equilibrium. In this case, specification of a certain number of variables (two of three variables—temperature, pressure, and volume—plus the masses and identities of all independent chemical substances in the system) will specify the state of the system. In other words, specification of the state is a recipe that allows us to reproduce the system at any time. It is an observed fact that if the state of a system is specified, its properties are given. The properties of a system are of two kinds. *Extensive* properties, such as volume and energy, require for their definition specification of the thermodynamic state including the amounts of all substances. *Intensive* properties, such as density or viscosity, are fixed by giving less information; only the *relative* amounts of different substances are needed. For example, the density of a $1\,M$ NaCl solution is independent of the size of the sample, though it depends on temperature (T), pressure (P), and the concentration of NaCl.

Thermodynamics is usually concerned with changes between equilibrium states. Such changes may be *reversible* or *irreversible*, depending on whether or not the path from initial to final state leads through a succession of equilibrium states. If a change is reversible, the system always lies so close to equilibrium that the direction of change can be reversed by an infinitesimal change in the surroundings.

Heat (q): The energy transferred into or out of a system as a consequence of temperature differences.

Work (w): Any other exchange of energy between a system and its surroundings. It may include such cases as volume change against external pressure, changes in surface area against surface tension, electrical work, and so forth.

Internal energy (E): The energy within the system. In chemistry we usually consider only those kinds of energy that might be modified by chemical processes. Thus energy involved in holding together the atomic nuclei is generally not counted. The internal energy of a system may then be taken to include the following: translational energy of the molecules, vibrational energy of the molecules, rotational energy of the molecules, the energy involved in chemical bonding, and, the energy involved in nonbonding interactions between molecules. Some such interactions are listed in Table 1.1.

The internal energy is a function of the state of a system. That is, if the state is specified, the internal energy is fixed at some value regardless of how the system came to be in that state. Since we are usually concerned with energy changes, internal energy is defined with respect to some arbitrarily chosen standard state.

Enthalpy $(H = E + PV)$: The internal energy of a system plus the product of its volume and the external pressure exerted on the system. It is also a function of state.

With these definitions, we state the first law of thermodynamics, an expression of the conservation of energy. For a change in state,

$$\Delta E = q - w \tag{1.1}$$

TABLE 1.1   NONCOVALENT INTERACTIONS BETWEEN MOLECULES

Type of Interaction	Equation[a]	Order of Magnitude[b] (kcal/mole)
Ion-ion	$E \sim \dfrac{z_1 z_2}{Dr}$	10–100
Ion-induced dipole	$E \sim \dfrac{z_1^2 \alpha_2}{Dr^4}$	1–10
Dipole-dipole	$E \sim \dfrac{\mu_1^2 \mu_2^2}{Dr^6 kT}$	0–1
Dipole-induced dipole	$E \sim \dfrac{\mu_1^2 \alpha_2}{D^2 r^6}$	0–1
Dispersion	$E \sim \dfrac{\alpha_1 \alpha_2}{r^6}$	1–10

[a] The subscripts 1 and 2 refer to the two molecules in the pair. Proportionality constants are not given. Symbols: z = charge, μ = dipole moment, α = polarizability, D = dielectric constant, r = distance between molecules, T = absolute temperature, and k is Boltzmann's constant.

[b] Compare to covalent bonds at 50–200 kcal/mole and hydrogen bonds at 1–10 kcal/mole. The values given are *very* approximate.

which takes the convention that heat absorbed by a system and work done by a system are positive quantities. For small changes, we write

$$dE = \dq - \dw \tag{1.2}$$

The slashes through the differential symbols remind us that while E is a function of state and dE is independent of the path of the change, q and w do depend on the path. The first law is entirely general and does not depend on assumptions of reversibility and the like.

If the only kind of work done involves change of the volume of the system against an external pressure ($P\,dV$ work),

$$dE = \dq - P\,dV \tag{1.3}$$

Similarly, we write for the change in the enthalpy of a system the general expression

$$dH = d(E + PV) = dE + P\,dV + V\,dP$$

$$= \dq - \dw + P\,dV + V\,dP \tag{1.4}$$

For systems doing only $P\,dV$-type work, $\dw = P\,dV$, and

$$dH = \dq + V\,dP \tag{1.5}$$

Equations (1.3) and (1.5) point up the meaning of dE and dH in terms of measurable quantities. For changes at constant volume

or
$$dE = dq$$
$$\Delta E = q_v \tag{1.6}$$

whereas, for processes occurring at constant pressure, (1.5) gives

$$dH = dq$$
$$\Delta H = q_p \tag{1.7}$$

That is, the heat absorbed by a process at constant volume measures ΔE, and the heat absorbed by a process at constant pressure measures ΔH. These quantities of heat will in general differ, because in a change at constant pressure some energy exchange will be involved in the work done in the change of volume of the system.

The thermochemistry of biological systems is almost always concerned with ΔH, since most natural biochemical processes occur under conditions more nearly approaching constant pressure than constant volume. However, since most such processes occur in liquids or solids rather than in gases, the volume changes are small. To a good approximation, one can often neglect the differences between ΔH and ΔE in biochemistry and simply talk about the "energy change" accompanying a given reaction.

Figure 1.1 summarizes compactly the relationships between the quantities q, w, ΔE, and ΔH. Note that we begin with the perfectly general first law and specialize to particular kinds of processes by adding more and more restrictions.

1.2 THE MOLECULAR INTERPRETATION OF THERMODYNAMIC QUANTITIES

We have seen that from the first law, together with the assertion that the internal energy is a function of the state of a system, powerful and general conclusions can be drawn. While these have not required molecular models to attest their validity, the student should keep in mind that quantities such as the internal energy and the energy changes in chemical reactions are ultimately expressible in terms of the behavior of atoms and molecules. It will be worth our while to explore the point in more detail.

Suppose we ask the following question: If we put energy into a system, to give an increase in the internal energy, where has the energy gone? Surely some has appeared as increased kinetic energy, but if the molecules are complex, some must be stored in rotational and vibrational energy and in intermolecular interactions, and perhaps some is accounted for by excited electronic states of a few molecules. Therefore, the question is really one of how the energy is *distributed*.

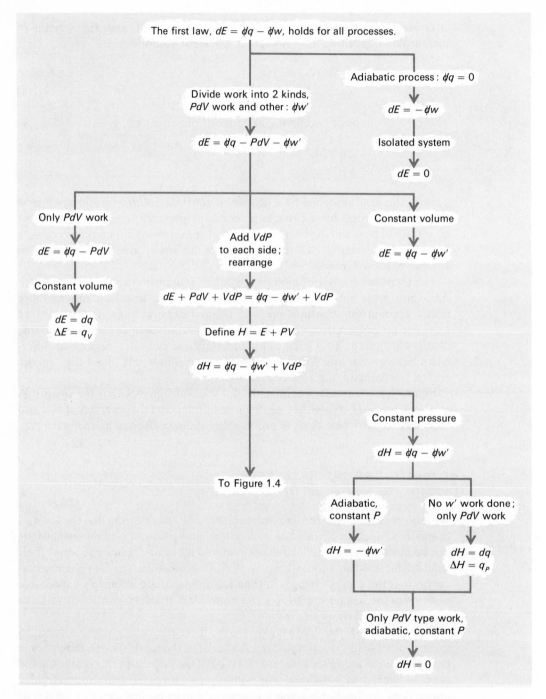

Figure 1.1   Consequences of the first law and the definition of ΔH. The "other" kind of work, $đw'$, may be identified with electrical work, work done in expanding a surface, and so forth.

For thermodynamic properties, we are going to be talking about large numbers of molecules, generally in or near states of equilibrium. The first means that a *statistical* point of view may be taken; we need not follow the behavior of any one molecule. The second implies that we should look for the *most probable* distribution of energy, for we would not expect an equilibrium state to be an improbable one. While any system might, by momentary fluctuations, occasionally distribute its energy in some improbable way (like having almost all of the energy in a few molecules), the relative occurrence of such extreme fluctuations becomes vanishingly small as the number of molecules becomes very large.

To see the principles involved, let us take a very simple system, a collection of particles that might be thought of, for example, as atoms in a gas or as protein molecules in a solution. Each of these entities is assumed to have a set of energy states available to it, as shown in Figure 1.2. The energy states available to a particle are not to be confused with the thermodynamic "states" of a system of many particles. Rather, they are the quantized states of energy accessible to any particle under the constraints to which the whole collection of particles is subject. Suppose we have six particles and a total energy of 10ϵ. Some distributions are shown in Figure 1.2, each of which satisfies the total energy requirement. Now let us say that for any particle, any state is equally probable.

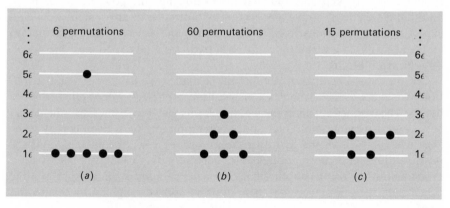

Figure 1.2   Some distributions of particles over energy states, subject to the constraints that $N = 6$ and $E = 10\epsilon$. The numbers are calculated from Equation (1.8) (remember $0! = 1$). The Boltzmann distribution most closely resembles (b). Of course, there are not enough particles for it to hold accurately in this simple case.

This simply means that there is nothing to prejudice a particle to pick a given state. *Then the most probable distribution will be the one that corresponds to the largest number of ways of rearranging particles over the states.* If we label the particles, we see that there are only six ways of making (a) and many more ways of making either (b) or (c). The number of ways of arranging N particles, n_1, in one group, n_2 in another, and so forth, is

$$W = \frac{N!}{n_1!n_2!n_3!\cdots n_i!\cdots} \qquad (1.8)$$

Since the most probable distribution is the one that corresponds to the largest number of arrangements of particles over energy states, the problem of finding that distribution is a problem of maximizing the number W, subject to the restrictions that N and total energy E are constants. Using standard mathematical techniques for handling such problems, the result for a large number of particles is found to be

$$n_i = n_1 e^{-a(\varepsilon_i - \varepsilon_1)} \qquad (1.9)$$

where n_i is the number of particles in energy state i, n_1 is the number in the lowest state, and ε_i and ε_1 are the energies of states i and 1, respectively. The constant, a, turns out to be $1/kT$, where k is the so-called Boltzmann constant, the gas constant R divided by Avogadro's number, and T is the absolute temperature. A simple derivation is given in W. J. Moore's *Physical Chemistry*, Chapter 15 (1963). For a more leisurely, but very clear discussion, see R. W. Gurney (1949).

Equation (1.9) is referred to as the Boltzmann distribution of energies. It should always be kept in mind that this is not the only possible distribution, and that if we could sample a collection of molecules at any instant we would expect to find deviations from it. It is simply the *most probable* distribution and hence will serve well if the number of particles is large and the system is at, or near, equilibrium.

One more modification of (1.9) should be made. We have written a distribution over energy *states*, whereas a distribution over *levels* would often be more useful. The distinction lies in the fact that levels may be degenerate—there may be several atomic or molecular states corresponding to a given energy level. (The energy levels of the hydrogen atom will serve as one example and the possible different conformational states corresponding to a given energy for a random-coil polymer as another.) If each level contains g_i states (that is, if the degeneracy is some integer g_i), then levels should be weighted by this factor. Then

$$n_i = \left(\frac{g_i}{g_1}\right) n_1 e^{-(\varepsilon_i - \varepsilon_1)/kT} \qquad (1.10)$$

where n_i and n_1 now refer to the number of particles in energy levels i and 1, respectively.

Equation (1.10) states that if the degeneracies of all states are equal, the lowest states will be the most populated at any temperature. At $T = 0$, $n_i = 0$ for $i > 1$, which means that all particles will be in the lowest level, while as $T \longrightarrow \infty$, the distribution tends to become more and more uniform. At high temperatures, no level is favored over any other, except for the factor of degeneracy. Another useful form of Equation (1.10) involves N, the total number of particles, instead of n_1. If we recognize that $N = \Sigma_i \, n_i$ (the sum being taken over all levels), then

$$N = \left(\frac{n_1}{g_1}\right) \sum_i g_i e^{-(\varepsilon_i - \varepsilon_1)/kT}$$

or

$$\frac{n_i}{N} = \frac{n_1}{g_1} \frac{g_i e^{-(\varepsilon_i - \varepsilon_1)/kT}}{\left(\dfrac{n_1}{g_1}\right) \sum_i g_i e^{-(\varepsilon_i - \varepsilon_1)/kT}}$$

$$= \frac{g_i e^{-(\varepsilon_i - \varepsilon_1)/kT}}{\sum_i g_i e^{-(\varepsilon_i - \varepsilon_1)/kT}} \tag{1.11}$$

The sum in Equation (1.11) is frequently encountered in statistical mechanics. It is sometimes called (for obvious reasons) the *sum over states* and more often (for less obvious reasons) the *molecular partition function*.

Since Equation (1.11) gives the fraction of molecules with energy ε_i, it is very useful for calculating average quantities. We shall make use of this idea in subsequent chapters. For example, in Chapter 6 we shall calculate the average contribution of dipolar molecules to solution polarizability in just this way.

1.3 ENTROPY, FREE ENERGY, AND EQUILIBRIUM

Second Law of Thermodynamics

So far, in discussing chemical and physical processes, we have concentrated on the energetics. We have shown that the first law, a restatement of the conservation of energy, could lead to exceedingly useful and general conclusions about the energy changes that accompany these processes. But one factor has been pointedly omitted—there has been no attempt to predict the *direction* in which changes will occur. Thus the first law allows us to discuss the heat transfer and work accompanying a chemical reaction such as

$$H_2O + ATP \longrightarrow ADP + \text{inorganic phosphate}$$

but gives no indication as to whether or not ATP will spontaneously hydrolyze in aqueous solution.

Intuitively, we would expect that under given conditions some particular equilibrium will exist between H_2O, ATP, ADP, and inorganic phosphate, but there is no way in which the first law can tell us where that equilibrium lies.

As another example, consider the dialysis experiment shown in Figure 1.3. A solution of sucrose has been placed in a dialysis bag immersed in a container

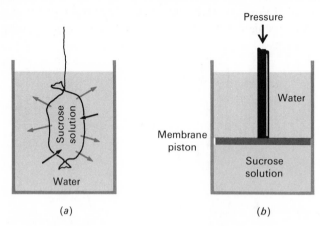

(a)   *(b)*

Figure 1.3   *(a)* A dialysis experiment in which sucrose will diffuse out of a bag and water in until equilibrium is attained. This process is irreversible. No work is done. *(b)* A way of doing the same experiment reversibly. The membrane piston is impermeable to sucrose and permeable to water. If the pressure on the piston is gradually reduced, the same final state (uniform mixing) approached irreversibily in *(a)* will be approached reversibly. With this arrangement, work will be done.

of water. We shall assume that the membrane is permeable to sucrose molecules. Either intuition or a simple experiment will tell us that the system is not at equilibrium. Rather, we know and find that sucrose will diffuse through the membrane until the concentrations inside and outside the bag are equal.

What is it that determines this position of equilibrium and the spontaneous (irreversible) process that leads to it?

A first guess might be made from analogy to mechanical processes. A mechanical system reaches a state of equilibrium when the energy of the system is a minimum. (Think of a ball rolling to the bottom of a hill.) But this will clearly not do here. Dilute sucrose solutions are close to ideal (see Chapter 2), which means that the energy of the system is practically independent of concentration. In fact, the interaction of sucrose with water is such that the state of lower concentration is actually of higher energy.

A moment's thought shows that the equilibrium state is favored in this case because of its *higher probability*. If one imagines a vast array of such systems, each of which has been left to itself for a considerable period, it is clear that the great majority will have the sucrose distributed quite uniformly throughout. True, there might be a very, very rare occurrence in which the sucrose concentration was appreciably higher inside or outside. But such systems will be very exceptional, for they would require the remarkable coincidence that a large excess of randomly wandering sucrose molecules happened to be inside or outside of the bag. A closer inspection shows that the main difference between the initial and final states of the system lies in the number of ways (W) in which sucrose molecules can be distributed over the total volume available to them. There are more ways of putting N sucrose molecules into a large volume than into a small volume. Thus one of the determinants of the equilibrium state of a system of many particles is the *randomness* of the system, or the number of ways (W) in which the particles of the system may be distributed, whether it be over levels of energy or, as in this instance, over the volume of the system.

For this reason, we define a function of state, which we shall call the *entropy*:

$$S = k \ln W \tag{1.12}$$

where k is the Boltzmann constant. The logarithm of W is chosen for the following reason: We wish the entropy to be an extensive property; that is, we wish the S of a system that is made up of two parts (1 and 2) to be $S_1 + S_2$. Now if W_1 is the number of ways of distributing the particles in part 1 in the particular state and W_2 the corresponding quantity for part 2, the number of ways for the whole system in this state is $W_1 W_2$:

$$S = k \ln W = k \ln W_1 W_2 = k \ln W_1 + k \ln W_2 = S_1 + S_2 \tag{1.13}$$

The above description of S, while exceedingly useful for certain problems, is not clearly related to measurable physical quantities. This may be accomplished in the following way: Let us assume that the molecules in a system have available a certain set of energy states and are distributed over these states according to the Boltzmann distribution

$$n_i = n_1 e^{-(\varepsilon_i - \varepsilon_1)/kT} \tag{1.14}$$

Now

$$W = \frac{N!}{n_1! n_2! n_3! \ldots n_n!} \tag{1.15}$$

$$\ln W = \ln(N!) - \ln(n_1!) - \ln(n_2!) - \cdots \tag{1.16}$$

or, using Stirling's approximation ($\ln n! \cong n \ln n - n$),

$$\ln W \cong N \ln N - N - n_1 \ln n_1 + n_1 - n_2 \ln n_2 + n_2 - \cdots \tag{1.17}$$

Since $\Sigma_i \, n_i = N$, we obtain

$$\ln W = K - \sum_i n_i \ln n_i \tag{1.18}$$

where K is the constant $N \ln N$.

Let us consider an infinitesimal change in the state of the system. This is assumed to be a *reversible* change; that is, the system does not depart appreciably from a state of equilibrium but simply changes from the Boltzmann distribution appropriate to the initial state to that appropriate to the infinitesimally different new state:

$$dS = kd \ln W \tag{1.19}$$

$$= -k \, d \sum_i n_i \ln n_i = -k \sum_i n_i \, d \ln n_i - k \sum_i \ln n_i \, dn_i \tag{1.20}$$

$$= -k \sum_i \ln n_i \, dn_i \tag{1.21}$$

The first term in (1.20) vanishes because it equals $\Sigma_i \, dn_i = d \, \Sigma_i \, n_i$ and the total number of particles is constant. The expression for dS can then be evaluated by remembering that $n_i = n_1 e^{-(\varepsilon_i - \varepsilon_1)/kT}$, so that

$$\ln n_i = \ln n_1 - \Delta \varepsilon_i / kT \tag{1.22}$$

where $\Delta \varepsilon_i$ is used to abbreviate $(\varepsilon_i - \varepsilon_1)$. Again dropping a term that is the sum of dn_i, we obtain

$$dS = k \left(\frac{\sum \Delta \varepsilon_i}{kT} \right) dn_i = \frac{1}{T} \sum_i \Delta \varepsilon_i \, dn_i \tag{1.23}$$

Now the sum in (1.23) is simply the heat absorbed in the process, for if a system has a given set of *fixed* energy levels and a reversible change in the population of the levels occurs, this must involve the absorption or release of heat from the system:

$$dS = \frac{dq_{rev}}{T} \tag{1.24}$$

This is the classic definition of an entropy change. For a finite isothermal process, we say

$$\Delta S = \frac{q_{rev}}{T} \tag{1.25}$$

The heat, q_{rev}, is a perfectly defined quantity, since the change is required to be reversible.

Equation (1.25) states that if we wish to calculate the entropy change in the transition between state 1 and state 2 of a system, it is only necessary to consider a reversible path between the two states and calculate the heat absorbed or evolved. As a simple example, we may consider the case of an isothermal, reversible expansion of 1 mole of an ideal gas from volume V_1 to volume V_2. For each infinitesimal part of the process, $dE = 0$ (since the gas is ideal and its energy is independent of its volume). Then

$$dq = dw = P\,dV = \frac{RT\,dV}{V} \tag{1.26}$$

and

$$\frac{q_{rev}}{T} = \frac{RT}{T}\int_{V_1}^{V_2}\frac{dV}{V} = R\ln\!\left(\frac{V_2}{V_1}\right) \tag{1.27}$$

so

$$\Delta S = R\ln\!\left(\frac{V_2}{V_1}\right) \tag{1.28}$$

The entropy of the gas increases in such an expansion.

We can arrive at exactly the same result from the statistical definition of entropy. Suppose we divide the volume V_1 into n_1 cells, each of volume V: therefore $V_1 = n_1 V$. The larger volume V_2 is divided into n_2 cells of the same size: $V_2 = n_2 V$. Putting one molecule into the initial system of volume V_1, there are n_1 ways in which this can be done. For two molecules there will be n_1^2 ways. For an Avogadro's number of molecules, the number of ways to put them into V_1 is $W = n_1^{\mathcal{N}}$. If the larger volume V_2, containing V_2/V cells, is occupied by the same $\mathcal{N}$ molecules, $W_2 = n_2^{\mathcal{N}}$. From Equation (1.12),

$$\Delta S = k\ln n_2^{\mathcal{N}} - k\ln n_1^{\mathcal{N}}$$

$$= k\ln(n_2/n_1)^{\mathcal{N}} = R\ln(n_2/n_1) = R\ln\!\left(\frac{V_2/V}{V_1/V}\right) \tag{1.29}$$

where we have made use of the fact that $\mathcal{N}k = R$. Then,

$$\Delta S = R\ln V_2/V_1 \tag{1.30}$$

Note that the cell volume V has canceled out of the final result; it is just a device to allow us to compare the number of ways of making the systems 1 and 2. Note also that it was necessary for the constant in Equation (1.12) to have the value k (Boltzmann's constant) in order that the two methods of approach would lead to the same numerical value of ΔS.

A very similar calculation could be carried out for our example of sucrose diffusing out of a dialysis bag. In this case, however, the calculation from q_{rev}/T would be more difficult, for we would have to imagine some reversible way in which to carry out the process. One such way is depicted in Figure 1.3(*b*). The dialysis bag is replaced by a membrane piston, to which is exerted a pressure equal to the osmotic pressure of the solution (see Chapter 2). If this pressure is gradually reduced, the piston will rise, doing work on the surroundings and diluting the sucrose solution. Since once again the ideality of the system requires $\Delta E = 0$, heat must be absorbed to keep the system isothermal.

Both the classic and statistical calculation will again lead to the same result, which is formally identical to that for the gas expansion. Per mole, we have

$$\Delta S = R \ln \frac{V_2}{V_1} = R \ln \frac{C_1}{C_2} \tag{1.31}$$

where C_1 and C_2 are concentrations.

A very similar calculation allows us to determine the entropy of mixing for an ideal solution made up of N components. Assume that there are N_i molecules of each component, making a total number of N_0 molecules in the whole solution: $N_0 = \sum_{i=1}^{N} N_i$. We assume that the molecules are about the same size, so that each can occupy a "cell" of the same volume. Since the mixture is assumed to be ideal, there will be no volume change in mixing, and the total number of cells in the solution will be N_0, the total number of molecules. The entropy of mixing is defined as

$$\Delta S_m = S(\text{solution}) - S(\text{pure components}) \tag{1.32}$$

We are concerned only with mixing entropy, so we may say that each of these quantities is given by $k \ln W$, where W is the number of distinguishable ways of putting molecules in cells. So

$$\Delta S_m = k \ln \frac{W(\text{solution})}{W(\text{pure components})} \tag{1.33}$$

But all arrangements of molecules in the pure components are the same, so $W(\text{pure components}) = 1$, and $\Delta S_m = k \ln W(\text{solution})$. But $W(\text{solution})$ is just the number of ways of arranging N_0 cells into groups with N_1 of one type, N_2 of another, and so forth. Therefore we have the familiar expression

$$\Delta S_m = k \ln \frac{N_0!}{N_1! N_2! \cdots N_N!} \tag{1.34}$$

Since all of the N's are large, we may use Stirling's approximation:

$$\Delta S_m \cong k\{N_0 \ln N_0 - N_0 - N_1 \ln N_1 \\ + N_1 - N_2 \ln N_2 + N_2 - \cdots\} \tag{1.35}$$

or, since

$$\sum_i N_i = N_0,$$

$$\Delta S_m = k\{N_0 \ln N_0 - \sum_i N_i \ln N_i\} \tag{1.36}$$

or

$$\Delta S_m = k\{\sum_i N_i \ln N_0 - \sum_i N_i \ln N_i\} \tag{1.37}$$

$$= -k \sum_i N_i \ln \frac{N_i}{N_0} = -k \sum_i N_i \ln X_i \tag{1.38}$$

where X_i is the mole fraction of i. Multiplying and dividing by Avogadro's number, we obtain the final result:

$$\Delta S_m = -R \sum_i n_i \ln X_i \tag{1.39}$$

where n_i is the number of moles of component i. This expression says that even if there is no interaction between the molecules, the entropy of a mixture is always greater than that of the pure components, since all X_i's are less than unity.

It is evident from the above examples and from Equation (1.12) that the entropy can be considered as a measure of the randomness of a system. A crystal is a very regular and nonrandom structure; the liquid to which it may be melted is much more random and has a higher entropy. The entropy change in melting can be calculated as

$$\Delta S = \frac{q_{rev}}{T} = \frac{\Delta H_{melting}}{T_{melting}} \tag{1.40}$$

Since the melting of a crystalline solid will always be an endothermic process and since T is always positive, the entropy will always increase when a solid is melted. Similarly, a *native* protein molecule is a highly organized, regular structure. When it is *denatured*, an unfolding and unraveling takes place; this corresponds to an increase in entropy. Of course, this is not the whole story; upon denaturation the interaction of solvent with the protein may change in such a way either to add or subtract from this entropy change.

Whenever a substance is heated, the entropy will increase. We can see this in the following way: At low temperatures, only the few lowest of the energy levels available to the molecules are occupied; there are not too many ways to do this. As T increases, more levels become available and the randomness of the system increases. The entropy change can, of course, be calculated by assuring that the heating is done in a reversible fashion. Then, if the process is at constant pressure, $dq_{rev} = C_P \, dT$, where C_P is the constant pressure heat capacity

$$dS = \frac{C_P \, dT}{T} \tag{1.41}$$

or, in heating a substance from temperature T_1 to T_2,

$$\Delta S = \int_{T_1}^{T_2} \frac{C_P}{T}\, dT \tag{1.42}$$

If C_P is constant,

$$\Delta S = C_P \ln\left(\frac{T_2}{T_1}\right) \tag{1.43}$$

The definition of the entropy change in an infinitesimal, reversible process as dq_{rev}/T allows us a reformulation of the first law for reversible processes, since

$$dE = dq - dw = T\, dS - P\, dV \tag{1.44}$$

This leads to one precise answer to a question posed at the beginning of this section: What are the criteria for a system to be at equilibrium? If, and only if, a system is at equilibrium, an infinitesimal change will be reversible. We may write from Equation (1.44) that if the volume of the system and its energy are held constant,

$$dS = 0 \tag{1.45}$$

for a reversible change. This means that S must be either at a maximum or minimum for a system at constant E and V to be at equilibrium. Closer analysis reveals that it is the former. *If we isolate a system* (keep E and V constant), then it will be at equilibrium only when the entropy reaches a maximum. In any nonequilibrium condition the entropy will be spontaneously increasing toward this maximum. This is equivalent to the statement that an isolated system will approach a state of maximum randomness. This is the most common way of stating the *second law of thermodynamics.*

Isolated systems kept at constant volume are not of great interest to biochemists. Most of our experiments are carried out under conditions of constant temperature and pressure. To elucidate the requirements for equilibrium in such circumstances, a new function must be defined. This is the *Gibbs free energy, G*:

$$G = H - TS \tag{1.46}$$

We obtain the general expression for a change dG by differentiating (1.46):

$$dG = dH - T\, dS - S\, dT$$

$$= dE + V\, dP + P\, dV - S\, dT - T\, dS \tag{1.47}$$

This collection of terms simplifies if we note that $dE = T\, dS - P\, dV$ if the

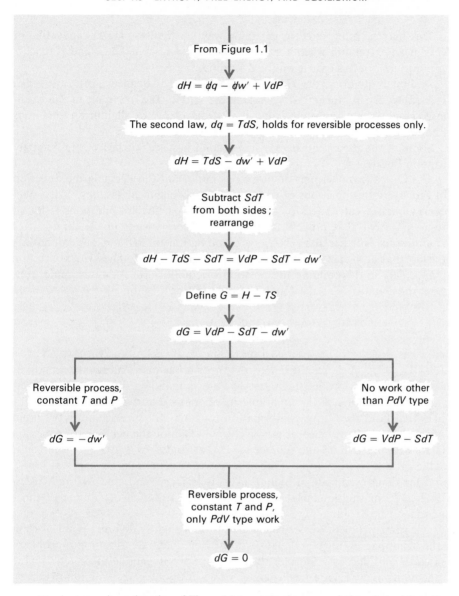

From Figure 1.1

$$dH = \not{d}q - \not{d}w' + VdP$$

The second law, $dq = TdS$, holds for reversible processes only.

$$dH = TdS - dw' + VdP$$

Subtract SdT
from both sides;
rearrange

$$dH - TdS - SdT = VdP - SdT - dw'$$

Define $G = H - TS$

$$dG = VdP - SdT - dw'$$

Reversible process,
constant T and P

No work other
than PdV type

$$dG = -dw' \qquad\qquad\qquad\qquad dG = VdP - SdT$$

Reversible process,
constant T and P,
only PdV type work

$$dG = 0$$

Figure 1.4   A continuation of Figure 1.1 to summarize some relations derived from the second law. Note that here q, w, and w' are those for reversible processes only.

process causing dG is reversible. Then

$$dG = V\,dP - S\,dT \qquad \text{reversible process} \tag{1.48}$$

And now, if the system is also constrained to constant T and P,

$$dG = 0 \qquad \text{reversible process} \tag{1.49}$$

This may be interpreted in the same way as Equation (1.45). The value of G is at an extremum when a system constrained to constant T and P reaches equilibrium. In this case it will be a minimum.

A similar quantity, the *Helmholtz free energy*, is defined as $A = E - TS$. It is of less use in biochemistry, because $dA = 0$ for the less frequently encountered conditions of constant T and V. By *free energy* we shall mean the Gibbs free energy.

Some of these consequences of the second law are worked out in compact form in Figure 1.4.

The Gibbs free energy is of enormous importance in deciding the direction of processes and positions of equilibrium in biochemical systems. If the free-energy change calculated for a process under particular conditions is found to be negative, that process is spontaneous, for it leads in the direction of equilibrium. Furthermore, ΔG, as a combination of ΔH and ΔS, emphasizes the fact that both energy minimization and entropy (randomness) maximization play a part in determining the position of equilibrium. As an example, consider the denaturation of a protein or polypeptide:

$$\Delta G_{den} = \Delta H_{den} - T\Delta S_{den} \tag{1.50}$$

Now we expect ΔS_{den} to be a positive quantity. By definition, we have $\Delta S_{den} = R \ln(W_{den}/W_{native})$. Since the denatured protein has many more conformations available to it than does the native, it follows that $W_{den}/W_{native} \gg 1$. Furthermore, the breaking of the favorable interactions (hydrogen bonding and the like) that hold the native conformation together will surely require the input of energy, so ΔH will also be positive. If we examine the behavior of Equation (1.50) with positive ΔS and ΔH, we see that at some low T, $\Delta G > 0$, whereas at sufficiently high T, $\Delta G < 0$. Thus the native state should be stable at low T and the denatured state at high T. This is in fact found, and ΔH and ΔS estimated from simple considerations are of the order of magnitude of those observed (see Problem 4).

The same problem can be considered from a slightly different point of view, which emphasizes the close relation of the Boltzmann distribution to problems of order and energy. Consider the energy of a polypeptide to be represented schematically by Figure 1.5. There is, we assume, one energy minimum corresponding to the ordered conformation and a host of higher energy conformations corresponding to the random-coil form. Of course, this is a gross oversimplification, and it must be understood that Figure 1.5 is a schematic representation of a multidimensional energy surface. We may write the Boltzmann equation for the ratio of numbers of molecules in the two forms as

$$\frac{n_{den}}{n_{nat}} = \frac{g_d}{g_n} e^{-(\varepsilon_d - \varepsilon_n)/kT} \tag{1.51}$$

We can then identify the degeneracies of these levels with the number of

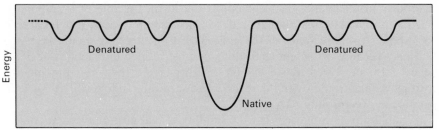

Figure 1.5 A *highly* schematic representation of the energy states of a protein. We assume that the molecule has only a single native state and a large number of denatured conformations of equal energy. These assumptions are certainly far too severe, but they allow a simple calculation. See the text.

conformations corresponding to each:

$$g_n = W_n = 1 \tag{1.52}$$

$$g_d = W_d \gg 1 \tag{1.53}$$

Now, we can rewrite (1.51) as

$$\frac{n_d}{n_n} = e^{\ln(W_d/W_n)} e^{-(\varepsilon_d - \varepsilon_n)/kT}$$

$$= e^{-[(\varepsilon_d - \varepsilon_n) - kT \ln(W_d/W_n)]/kT} \tag{1.54}$$

or, putting quantities on a molar basis,

$$\frac{n_d}{n_n} = e^{-[(E_d - E_n) - RT \ln(W_d/W_n)]/RT} \tag{1.55}$$

Both sides of this equation are recognizable. In the exponent, if we neglect the difference between ΔE and ΔH and note that the $\ln W$ term is an entropy change, we have

$$\frac{n_d}{n_n} = e^{-(\Delta H - T\Delta S)/RT}$$

$$= e^{-\Delta G/RT} \tag{1.56}$$

The left side is simply the equilibrium constant for the reaction. Therefore,

$$\ln K = \ln\left(\frac{n_d}{n_n}\right) = -\Delta G/RT \tag{1.57}$$

This equation will be familiar to most readers, and we shall derive it more rigorously in Chapter 3. Here, it points out again the role of randomness: The denatured state may be favored, in some circumstances, just because it corresponds to more states of the molecules. Conversely, it should be noted that only because of the energetically favorable residue interactions does the helical form described above exhibit stability. Were these not present, the entropy change would drive the reactions toward the random coil at *any* temperature above absolute zero.

It should not be supposed that the stability of real macromolecular conformations is as simple as implied above. For one thing, the solvent cannot, in general, be neglected, and solute-solvent interactions may play a very important role. To take a particular example, it has been argued that the stability of some macromolecules may derive in part from *hydrophobic bonding* (see Barker, in this series, Chapter 3). Such bonding may lead to a very different temperature dependence of stability. In the breaking of a hydrophobic bond, nonpolar groups are separated from one another and put in contact with the solvent. In aqueous solution, such groups are expected to become surrounded by shells of "icelike" water. The immobilization of a large number of water molecules should correspond to an entropy decrease—perhaps so much as to override the entropy increase accompanying the macromolecule's gain of conformational freedom. If this is the case, the overall entropy change for the transition from the ordered conformation of the macromolecule to the random one would involve a *decrease* in entropy. Then the temperature dependence should be a reverse of the above example; *low* temperatures should favor disorganization. Such behavior may be being observed when we see that some multisubunit proteins dissociate into their individual subunits at temperatures around 0°C.

PROBLEMS

Note: The problems in this chapter are intended mainly for review of the student's previous experience with physical chemistry. They may draw on details that have not been mentioned in this chapter.

1. Calculate q, w, ΔE, and ΔH for the isothermal expansion of 2 moles of an ideal gas from 16 to 95 liters at 0°C.

2. The standard-state enthalpy change for the oxidation of palmitic acid

$$CH_3(CH_2)_{14}COOH(s) + 24O_2(g) \longrightarrow 17CO_2(g) + 16H_2O(l)$$

is −2,380 kcal.
  (*a*) Calculate ΔE.
  (*b*) Calculate the work done when 1 mole is oxidized at 1 atm pressure.

3. The heat of melting of ice at 1 atm and 0°C is +1.4363 kcal/mole. The density of ice under these conditions is 0.917 g/cm³ and the density of water 0.9998 g/cm³.
  (*a*) Calculate ΔE.

(*b*) Calculate ΔS.

*(*c*) What will the melting point be at 10 atm? Derive necessary equations, make necessary assumptions.

*4. Assuming that a polypeptide chain has only one α-helical conformation and that there are three possible orientations for each amino acid residue in the random-coil state, calculate ΔS for the conformational change

$$\alpha \text{ helix} \rightleftharpoons \text{random coil}$$

for a polypeptide of 100 residues. What value of ΔH per residue would be required to make the melting point (the temperature at which the equilibrium constant equals 1) be 50°C? Compare with the hydrogen-bond energy, estimated to be 0 to 3 kcal/mole.

5. Assuming (see Chapter 2) that the osmotic pressure is given by $\pi = RTC/M$, work out the ΔS for the process depicted in Figure 1.3. Assume that you start with 1 g of sucrose in 100 ml and increase the volume to 1 liter.

6. Derive expressions for the following:

$$\left(\frac{\partial H}{\partial P}\right)_{E,V}, \qquad \left(\frac{\partial G}{\partial T}\right)_{P}, \qquad \left(\frac{\partial G}{\partial P}\right)_{T}$$

7. Two energy levels of a molecule are separated by 1×10^{-15} ergs. The degeneracy of the higher level is twice that of the lower. Calculate
(*a*) The relative populations of these levels at 0°C.
(*b*) The temperature at which they will be equally populated.

* *Note:* In this and subsequent chapters the problems judged to be more difficult are marked with an asterisk.

REFERENCES

Thermodynamics and elementary statistical mechanics are well treated in any number of the better physical chemistry texts. I recommend especially:

Daniels, F. and R. A. Alberty: *Physical Chemistry*, 3rd ed., John Wiley & Sons, Inc., New York, 1966. A standard, well-balanced text.

Moore, W. J., *Physical Chemistry*, 3rd ed., Prentice-Hall, Inc., Englewood Cliffs, N.J., 1963. More detailed and rigorous than most of the available texts at this level.

In the realm of more specialized books, the following have been especially useful:

* Denbigh, K. G.: *The Principles of Chemical Equilibrium*, Cambridge University Press, London, 1955. A very different book—discursive, somewhat long, but beautifully clear.

* Gibbs, J. W.: *The Collected Works of J. W. Gibbs*, Yale University Press, New Haven, Conn., 1948. Not a useful text but worth examining to see a master at work.

* Gurney, R. W.: *Introduction to Statistical Mechanics*, McGraw-Hill Book Company, Inc., New York, 1949. An excellent book for the beginner in statistical thermodynamics.

* *Note:* In this and following chapters the starred references are available in paperback editions.

* Hill, T. L.: *Thermodynamics of Small Systems*, W. A. Benjamin, Inc., New York, 1963. The application of statistical methods to a number of kinds of problems of interest to the biochemist.

Kirkwood, J. G. and I. Oppenheim: *Chemical Thermodynamics*, McGraw-Hill Book Company, Inc., New York, 1961. A high-level, authoritative treatment; terse.

Klotz, I.: *Energetics in Biochemical Reactions*, Academic Press, Inc., New York, 1957. A very brief, nonmathematical introduction to thermodynamics for biochemists. Some nice examples and explanations.

SOLUTIONS OF

TWO | # MACROMOLECULES

The development of a true molecular biology was long inhibited by fundamental misunderstandings about the nature of solutions of proteins, carbohydrates, and other large molecules. In the early 1900s, recognition of some of the remarkable features of the behavior of such solutions led to the idea that these were not true solutions at all. Since they exhibited no measurable freezing-point depression and since the solutes could not easily be caused to crystallize and exhibited very slow diffusion, such solutions were called *colloids*. This carried the unfortunate connotation that the well-known physicochemical laws for "true" solutions should not apply and that the application of conventional solution thermodynamics in this field was of doubtful validity.

Of course, we now know that proteins, and high polymers in general, form true solutions. The freezing-point depression is indeed small and diffusion is slow, but these are only consequences of the high molecular weights of the solutes. Crystallization is more difficult but possible in almost every case. There is only one cause for reservation, and that is easily disposed of. Thermodynamic descriptions of solutions are dependent on there being a very large number of solute particles in the sample observed, so that macroscopic fluctuations in properties will be very unlikely. Is this condition satisfied? If we consider an extreme case, a solution containing 0.01 mg/ml of a virus of molecular weight 100 million, we shall still find roughly 10^{10} particles/ml. This is a number large enough that we need not worry about fluctuations. It is only when we begin considering volumes comparable to that of a single cell that such questions need be raised.

With this preface, we shall first review the ways in which the thermodynamics of solutions is described and then apply these concepts to the description of solutions of macromolecules.

2.1   SOME FUNDAMENTALS OF SOLUTION THERMODYNAMICS

A solution is a single-phase system containing more than one component. A *component* is an independently variable chemical substance. It should be noted that this definition of component is strict and that a solution will frequently contain fewer components than molecular species that might be present. This will occur whenever chemical equilibria exist in the solution. To take an example, consider a solution containing water, the protein hemoglobin (Hb), and dissolved oxygen. Since each hemoglobin molecule can bind one, two, three, or four oxygen molecules, there will be a number of molecular species potentially or actually present, namely H_2O, Hb, O_2, HbO_2, $Hb(O_2)_2$, $Hb(O_2)_3$, and $Hb(O_2)_4$. Yet if the binding reactions are in equilibrium, there are only three *independently variable* substances, or components. Specification of the solvent plus any two of the others would, via the equilibrium relationships, specify the rest. Thus the solution contains three and only three components. Specification of the amounts of three of the substances, together with the temperature and pressure, will completely define the state of the system. It is important to emphasize that such simplification is possible only if the system is in equilibrium. If, in the above case, the oxygenation or deoxygenation reactions were very slow, we might perturb the system (put in some more oxygen, for example) and observe the system in a nonequilibrium state. In this case, more than three components would be needed to describe the system.

The description of the state of a solution by stipulation of T, P, and the amounts of n components may be thought of as a recipe. It means that if, on two or more occasions, we fix these variables at the same values, we shall find exactly the same properties, both extensive and intensive. We may logically ask, then, how some extensive property, like the volume, will depend on the amounts of the various components. It would be extremely naive to assume simple additivity. That is, we would not expect that n_i moles of each component of molar volumes V_i would, when mixed, give a total volume $V = \sum_i n_i V_i$. In many cases volume changes on mixing can easily be observed. The true situation is easy to see; if we add a small amount of component i to a mixture, the change produced in an extensive property will depend not only on the amount and nature of the substance added but also on the composition of the mixture to which addition is made. This leads to the definition of *partial molar* and *partial specific* quantities.

Considering any extensive property, X, we define the partial molar quantity, $\overline{X}_i$, as

$$\overline{X}_i = \left(\frac{\partial X}{\partial n_i}\right)_{T,P,n_{j \neq i}} \tag{2.1}$$

where n_i is the number of moles of component i. The subscripts indicate that we hold T, P, and the amounts of all components except i constant. The partial specific quantity, $\bar{x}_i$, is given by

$$\bar{x}_i = \left(\frac{\partial X}{\partial g_i}\right)_{T,P,g_{j \neq i}} \tag{2.2}$$

where g_i represents grams of component i. The latter quantities are frequently used in macromolecular chemistry, since we often do not know the molecular weight and hence the number of moles in a sample.

To make these definitions concrete, let us consider the volume of a solution, V, as the extensive property. The partial molar volume ($\overline{V}_i$) and partial specific volume ($\bar{v}_i$) of component i are then given by

$$\overline{V}_i = \left(\frac{\partial V}{\partial n_i}\right)_{T,P,n_{j \neq i}} \tag{2.3}$$

and

$$\bar{v}_i = \left(\frac{\partial V}{\partial g_i}\right)_{T,P,g_{j \neq i}} = \frac{\overline{V}_i}{M_i} \tag{2.4}$$

respectively.

These partial quantities are intensive variables, independent of the size of the system. Thus $\overline{V}_i$ will depend on the composition of the mixture of which i is a part and on T and P but not on the amount of the solution. This leads to a very important result, which bears on the original question as to how *extensive* properties may be calculated. Suppose we imagine putting together a volume V of solution by starting from nothing and adding infinitesimal increments of the n components, always in the proportions to be found in the final mixture. That is, we make $dg_1 = g'_1\, d\lambda$, $dg_2 = g'_2\, d\lambda, \ldots, dg_i = g'_i\, d\lambda$, where $g'_1, g'_2, \ldots, g'_i$ are constants equal to the number of grams of each substance in the final mixture and λ is a dummy variable going from 0 to 1. With this method of assembly, the composition will always be the same, and the $\bar{v}$'s will be constant during construction of the solution. Now

$$dV = \bar{v}_1\, dg_1 + \bar{v}_2\, dg_2 + \cdots + \bar{v}_i\, dg_i + \cdots + \bar{v}_n\, dg_n$$

$$= g'_1\bar{v}_1\, d\lambda + g'_2\bar{v}_2\, d\lambda + \cdots + g'_i\bar{v}_i\, d\lambda + \cdots + g'_n\bar{v}_n\, d\lambda \tag{2.5}$$

Since both the g_i''s and the $\bar{v}_i$'s are constants, we may integrate easily:

$$V = (g_1'\bar{v}_1 + g_2'\bar{v}_2 + \cdots + g_i'\bar{v}_i + \cdots + g_n'\bar{v}_n) \int_0^1 d\lambda \qquad (2.6)$$

$$= \sum_{i=1}^{n} g_i'\bar{v}_i \qquad (2.7)$$

The total extensive quantity is the sum of the partial specific quantities each multiplied by the number of grams of the appropriate component. Similarly, with partial molar quantities

$$X = \sum_{i=1}^{n} n_i \bar{X}_i \qquad (2.8)$$

These are the appropriate summation rules for real solutions.

A quantity that will be of the utmost significance in many of our problems is the partial molar free energy. We shall again and again be concerned with equilibria of multicomponent systems at constant T and P. Hence the free energy of the solution will be the property of interest, and we must know the contribution of each component to the total free energy. The partial molar free energy is given the special symbol μ and is defined as

$$\bar{G}_i = \mu_i = \left(\frac{\partial G}{\partial n_i}\right)_{T,P,n_{j \neq i}} \qquad (2.9)$$

The total free energy of the solution is then

$$G = \sum_{i=1}^{n} n_i \mu_i \qquad (2.10)$$

The partial molar free energy is often called the *chemical potential*. This term is appropriate, for as we shall see, differences in μ may be regarded as the driving forces for such processes as chemical reactions or diffusion in which changes in the amounts of chemical substances occur.

The change in G for an infinitesimal, reversible change in the state of an open system (in which amounts of components can change) is given by the general formula

$$dG = \left(\frac{\partial G}{\partial T}\right)_{P,n_i} dT + \left(\frac{\partial G}{\partial P}\right)_{T,n_i} dP + \sum_i \left(\frac{\partial G}{\partial n_i}\right)_{T,P,n_j} dn_i \qquad (2.11)$$

or

$$dG = -S\,dT + V\,dP + \sum_i \mu_i\,dn_i \qquad (2.12)$$

or, at constant T and P,

$$dG = \sum_i \mu_i \, dn_i \tag{2.13}$$

From the above results we can obtain an additional important and useful theorem, the Gibbs–Duhem equation. If we differentiate (2.10) generally,

$$dG = \sum_i \mu_i \, dn_i + \sum_i n_i \, d\mu_i \tag{2.14}$$

Combining with (2.13), we have, at constant T and P,

$$\sum_i n_i \, d\mu_i = 0 \tag{2.15}$$

This means that variations of μ_i are not independent of one another and that we can express the chemical potential of one component in terms of those of the others.

The importance of the chemical potential arises from the fact that it measures that increment of free energy accompanying an infinitesimal change in the amount of one particular component in a system. Thus it is immediately applicable to discussions of phase equilibria and chemical equilibria. Two very general statements, on which most of our use of the chemical potential will be based, are given below:

1. If a number of phases are in equilibrium, the chemical potential of a given component will have the same value in all phases in which that component is present. While the general proof will not be given here, the principle involved is obvious: Transfer of an infinitesimal amount of a component between two phases (α and β) in equilibrium at constant T and P must involve a zero free-energy change; this can only be so if μ_i is the same in both phases, since, from (2.13)

$$dG = \mu_i^{\alpha} \, dn_i^{\alpha} + \mu_i^{\beta} \, dn_i^{\beta} \tag{2.16}$$

but since

$$dn_i^{\alpha} = -dn_i^{\beta} \qquad \text{and} \qquad dG = 0,$$

we have

$$0 = (\mu_i^{\alpha} - \mu_i^{\beta}) \, dn_i^{\alpha} \tag{2.17}$$

$$\mu_i^{\alpha} = \mu_i^{\beta} \tag{2.18}$$

This principle will be the starting point for discussion of such problems as membrane equilibria.

2. In the general chemical equilibrium,

$$aA + bB + \cdots \rightleftharpoons gG + hH + \cdots \tag{2.19}$$

where a, b, and so forth are stochiometric numbers, the chemical potentials of the components A, B, and so forth must obey the relationship

$$a\mu_A + b\mu_B + \cdots = g\mu_G + h\mu_H + \cdots \tag{2.20}$$

if the system is to be at chemical equilibrium.

This result can be demonstrated easily. Suppose that the system is kept at constant T and P but that an infinitesimal displacement from equilibrium occurs. Since the free energy must have been at a minimum if the system was initially at equilibrium, $dG = 0$. Furthermore, according to the stochiometric equation (2.19), the dn_i's are not independent but must be in the proportions

$$\frac{dn_A}{a} = \frac{dn_B}{b} = \cdots = -\frac{dn_G}{g} = -\frac{dn_H}{h} = \cdots \tag{2.21}$$

Now dG may be written as

$$dG = \mu_A \, dn_A + \mu_B \, dn_B + \cdots + \mu_G \, dn_G + \mu_H \, dn_H + \cdots \tag{2.22}$$

which gives

$$0 = a\mu_A \frac{dn_A}{a} + b\mu_B \frac{dn_B}{b} + \cdots + g\mu_G \frac{dn_G}{g} + h\mu_H \frac{dn_H}{h} + \cdots \tag{2.23}$$

or from (2.21)

$$a\mu_A + b\mu_B + \cdots = g\mu_G + h\mu_H + \cdots \tag{2.24}$$

These statements about the chemical potential are of importance because the chemical potential of a substance in a mixture depends on its concentration. It is this dependence that allows us to link the very general statements concerning equilibrium to specific experimental problems, where we observe variations in the concentrations of substances. The question then is, How does μ_i depend on the concentration of i? To answer this, we must inquire a bit more into the nature of solutions.

In physical chemistry, the distinction is made between *ideal* and *nonideal* solutions. This is useful, since ideal behavior leads to exceedingly simple laws, which serve as prototypes for the more complex laws describing real, nonideal solutions. The physical chemist customarily defines an ideal solution as one in which all components follow Raoult's law; that is, the vapor pressure of each component is strictly proportional to the mole fraction of that component over the entire concentration range. We shall adopt a somewhat different definition, which leads to the same result but enables us to see directly how the chemical potential depends on concentration. An ideal solution will be defined as one in which

1. The enthalpy change in mixing is zero

$$\Delta H_m = 0 \qquad (2.25)$$

2. The entropy change in mixing is given by the simple statistical rule (see Chapter 1)

$$\Delta S_m = -R \sum_{i=1}^{n} n_i \ln X_i \qquad (2.26)$$

where the X_i's are mole fractions.

This definition is easy to visualize on the molecular basis; it means that there is no difference in interaction *energy* between solute and solvent ($\Delta H_m = 0$) and that the entropy change arises only from the randomness produced by mixing. From (2.25) and (2.26), we have, for the free energy of mixing,

$$\Delta G_m = \Delta H_m - T \Delta S_m = RT \sum_i n_i \ln X_i \qquad (2.27)$$

for an ideal solution. But in general, the free energy of mixing is defined by

$$\Delta G_m = G_{(solution)} - \sum_i G_{i(pure\ components)} \qquad (2.28)$$

The free energy of a particular pure component will equal the number of moles of that component times a free energy per mole. We may regard this free energy per mole of a pure substance as a reference chemical potential, μ_i^0. Then from Equations (2.28) and (2.10)

$$\Delta G_m = \sum_i n_i \mu_i - \sum_i n_i \mu_i^0 \qquad (2.29)$$

or by (2.27)

$$\sum_i n_i(\mu_i - \mu_i^0) = RT \sum_i n_i \ln X_i \qquad (2.30)$$

Since the components of a solution are independently variable, we must have

$$\mu_i - \mu_i^0 = RT \ln X_i \qquad (2.31)$$

That is, the chemical potential will depend on the logarithm of the concentration. In the *standard state* or *reference state*, which has been chosen in this case to be pure component i, $\mu_i = \mu_i^0$. While the mole fraction scale is satisfactory for describing the behavior of the solvent in a solution of macromolecules, one usually does not employ this concentration scale for the solute. We are interested in solute components that are present only at very low concentrations, and therefore the pure component as a standard state is inconvenient. At low concentrations, a weight or molar concentration will be proportional to mole fraction. Since it is the logarithm of the concentration that occurs in (2.31), any proportionality constant can be absorbed in a redefined μ_i^0. Thus we may equally well write for the solute in dilute, ideal solutions

$$\mu_i = \mu_i^0 + RT \ln C_i \tag{2.32}$$

where C_i denotes grams per liter. The value of μ_i^0 is understood to depend on the concentration scale used. In this case μ_i^0 is the chemical potential of the ideal solute at unit concentration, for example, when $C_i = 1$ g/liter.

Nonideality can be taken care of in a purely formal way by inserting an activity coefficient. Thus the analog of Equation (2.32) for a nonideal solution can be written as

$$\mu_i = \mu_i^0 + RT \ln C_i y_i \tag{2.33}$$

where y_i is a function that describes all deviations from ideality. In general, y_i will be a function of T, P, and *all* of the solute concentrations. We expect that, in general, solutions will approach ideal behavior as $C \longrightarrow 0$; thus

$$\lim_{c \to 0} y_i = 1 \tag{2.34}$$

where C is the total concentration of *all* solutes in the solution.

In discussing some properties of macromolecular solutions (for example, osmotic pressure) we shall be interested primarily in the chemical potential of the solvent. For dilute solutions it is most convenient to take the standard state of the solvent as pure solvent and to use the mole fraction scale. Then, we write, in place of (2.31),

$$\mu_1 = \mu_1^0 + RT \ln X_1 f_1 \tag{2.35}$$

where f_1 is an activity coefficient on the mole fraction scale. But since $X_1 = 1 - X_2$ (where X_2 is the mole fraction solute) and $\ln(1 - X_2) \simeq -X_2 - X_2^2/2 \cdots$,

$$\mu_1 - \mu_1^0 \simeq -RT(X_2 + \cdots) + RT \ln f_1 \tag{2.36}$$

Now we may wish to express the solute concentration in terms of C_2 rather than X_2. For dilute solutions

$$X_2 \simeq \frac{C_2 V_1^0}{M_2} \tag{2.37}$$

where V_1^0 is the molar volume of pure solvent. Therefore,

$$\mu_1 - \mu_1^0 = -\frac{RTV_1^0 C_2}{M_2} + RT \ln f_1 \qquad (2.38)$$

It is clear that both $\ln f_1$ and its first derivative must approach zero as $C_2 \longrightarrow 0$. Furthermore, we shall (in most cases) expect f_1 to be less than unity (reduced solvent activity) at finite C_2. Therefore we might expect to be able to represent $\ln f_1$ by a negative power series in C_2, starting with C_2^2:

$$\ln f_1 = -(\alpha C_2^2 + \beta C_2^3 + \cdots) \qquad (2.39)$$

where the α, β, and so forth are unknown constants. This gives, when combined with (2.38),

$$\mu_1 - \mu_1^0 = -RTV_1^0 \left\{ \frac{C_2}{M_2} + BC_2^2 + \cdots \right\}$$

$$= -\frac{RTV_1^0 C_2}{M_2} \{ 1 + BM_2 C_2 + \cdots \} \qquad (2.40)$$

This is known as the virial expansion of the solvent chemical potential. The quantities $1/M_2$, B, and so forth are known as the first, second, and higher virial coefficients. The quantity B, the second virial coefficient, serves as a convenient measure of solution nonideality. If $BC_2 \simeq 0$, the higher terms in the virial expansion will presumably be even smaller, and we may write

$$\mu_1 = \mu_1^0 - \frac{RTV_1^0 C_2}{M_2} \qquad (2.41)$$

This conforms to our intuitive expectation that the chemical potential of the solvent should be reduced by adding a bit of solute.†

2.2   SOLUTIONS OF MACROMOLECULES

The solution behavior of biological macromolecules is determined principally by three factors: They are large, they may interact strongly with the solvent or one another, and they frequently carry electrical charges. We shall consider each of these sources of the nonideality, but first it is necessary to say something about the conformations that a macromolecule may exhibit in solution.

---

† Note that a very small second virial coefficient will exist even for an ideal solution if we retain the first term that was discarded in deriving (2.36). Work this out and compare it to the values in Table 2.1 to see that it is negligible.

The subject of macromolecular conformation has been treated in some detail by Barker, in this series. Here we shall simply enumerate some general categories.

Random coils: A linear polymer that exhibits little resistance to rotation about the bonds in the chain and in which there is little side-group interaction will tend to adopt the kind of random conformation depicted in Figure 2.1(*a*). There is, in such a case, no unique three-dimensional structure; rather, the molecule is continually being contorted by the impacts of surrounding solvent molecules. Thus we can speak only of *average* dimensions for such molecules. For example, we define the average radius of gyration in terms of the average squared distances($\overline{R_i^2}$) of the segments from the center of mass:

$$R_G^2 = \frac{\sum\limits_i \overline{R_i^2}}{N} \qquad (2.42)$$

where N is the number of segments.

A statistical analysis of the behavior of such a random coil shows that the average distribution of segments about the center of mass will be spherically symmetric, with an approximately Gaussian distribution of segment density about the center. The value of R_G gives a rough measure of the effective radius of such a molecule. For a true random coil, R_G will be proportional to the square root of the number of segments (see C. Tanford, 1961.)

Rodlike macromolecules: Some biopolymers, like polynucleotides and helical polypeptides, adopt a rodlike conformation in solution. Molecules of this type usually have a simple helical secondary structure. Their solution properties are largely dominated by the length of the rod, the diameter playing a lesser role [Figure 2.1(*b*)].

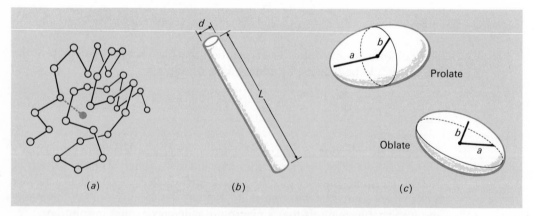

(a)   (b)   (c)

Figure 2.1   Three extreme classes of macromolecular conformation. (*a*) Random coil. The root-mean-square distance of the segments from the center of mass (black dot) determines the radius of gyration. (*b*) Rod, with length *L* and diameter *d*. (*c*) Prolate and oblate ellipsoids of revolution. The axial ratio is defined as *a/b*.

Globular macromolecules: Biopolymers in which there are strong side-chain interactions tend to coil up into dense globular conformations. Such are typical of the so-called globular proteins. Such molecules can often be adequately represented, for consideration of their solution properties, as either spheres or ellipsoids of low axial ratio. Since this kind of model will be used in a number of the following chapters, we shall describe both prolate (cigar-shaped) and oblate (flattened-sphere) ellipsoids in Figure 2.1(*c*).

It must not be assumed that such models exactly represent real macro-molecules. (For a comparison to a real protein molecule, see Chapter 11.) Nevertheless, for physical calculations, such approximate models frequently work very well.

We now return to a discussion of nonideal behavior in solutions of macro-molecules. In the first place, many macromolecules interact strongly with the solvent or with one another. This can have two effects: Either $\Delta H_m \neq 0$, or ΔS_m may contain contributions from ordering or disordering of the solvent. While quantitative analysis is difficult, it is not hard to present a qualitative picture of the effects, at least insofar as the ΔH_m contributions are concerned. A negative heat of mixing corresponds to favorable solute-solvent interaction.

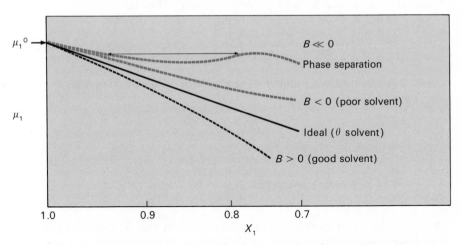

Figure 2.2   A schematic diagram of the behavior of the solvent chemical potential near $X_1 = 1$ for ideal and nonideal solutions. For $B \ll 0$, phase separation will occur, yielding the phases given by the ends of the arrow.

This should lead to an even greater decrease in the solvent chemical potential than expected for ideal solutions—and thus a positive value of B [see Equation (2.40), and Figure 2.2]. Conversely, if solute-solute interaction is strong, the virial coefficient will be negative. Too large a negative value of B is not to be tolerated, however, for if the value of B is so large that a minimum occurs in the μ_1 versus X_1 graph, phase separation must occur. (See the upper curve in Figure 2.2.) In this case there are two concentrations at which the solvent has the same chemical potential; this means that two liquid phases, of these two compositions, can coexist, for solvent can be transferred from one to the other with zero free-energy change.

The most general cause for nonideality in solutions of macromolecules arises from the sheer size of the particles. It will be recalled that an ideal solution is one for which the ideal entropy of mixing law holds. The derivation of that law (see Chapter 1) involved the assumption that the solute molecules are of the same order of size as the solvent molecules, so that solvent and solute might be interchanged at random in a hypothetical lattice. But this is by no means true for macromolecules. Each macromolecule is many times larger than a solvent molecule; in fact, the monomer units of the macromolecule are more comparable to the solvent molecules in size. In other words, a macromolecular solution more nearly resembles one in which the solute particles (the monomer units) are required to move together in clumps.

Put another way, this says that the distribution of solute molecules in a macromolecular solution can never be entirely random. The center of each molecule is excluded from a volume determined by the volumes occupied by all of the other molecules. It is not surprising, then, to find that the nonideality, and hence the virial coefficient, depend on the molecular excluded volume (u) [for a detailed derivation, see C. Tanford (1961)]. We find

$$B = \frac{\mathcal{N}u}{2M_2^2} \tag{2.43}$$

where $\mathcal{N}$ is Avogadro's number. The excluded volume is determined by the actual molecular dimensions.

$$\text{For spheres: } u = \frac{8M_2 v_2}{\mathcal{N}}; \qquad B = \frac{4v_2}{M_2} \tag{2.44}$$

$$\text{For rods: } u = \frac{2LM_2 v_2}{\mathcal{N}d}; \qquad B = \frac{L}{d}\frac{v_2}{M_2} \tag{2.45}$$

where v_2 is the specific volume, L the rod length, and d its diameter. In Table 2.1 are shown representative values of B and $(1 + BM_2 C_2)$ for $C_2 = 5$ mg/ml, a concentration in the range frequently employed in the study of macromolecules. The quantity $(1 + BM_2 C_2)$ represents the correction term in Equation (2.40). The general conclusion to be drawn is that while this contribution to non-ideality is small for spherical molecules, it can become large for very asymmetric rods or random coils. This is understandable, for the number of ways in which such particles can be packed into a solution is quite limited.

In view of the above, it might seem surprising that solutions of macromolecules can exhibit very nearly ideal behavior under some conditions. For example, for random-coil polymers there is often a particular temperature (called the θ temperature) at which $B = 0$. Qualitatively, the reason for this is the following.

TABLE 2.1   **SOLUTION NONIDEALITY**

Particle	B^a	$(1 + BM_2C_2)^b$
Sphere, $M = 10^5$	3×10^{-5}	1.015
Rod, $L/d = 100$, $M = 10^5$	7.5×10^{-4}	1.375
Random coil, $M = 10^5$, good solvent	5×10^{-4}	1.250
Random coil, θ solvent	0	1.000

^{a} Order of magnitude, in units of $cm^3 g^{-2} mole$.
^{b} Assuming that $C_2 = 5$ mg/ml.

The excluded volume effect is always such as to lower the chemical potential of the solvent. It is an entropy effect. On the other hand, in poor solvents the solute-solute interaction may be such as to increase the solvent chemical potential. This is an enthalpic term. There may exist a temperature (the θ temperature) at which these enthalpic and entropic terms exactly cancel. Then the solution will behave ideally. Putting this in mechanistic terms, we regard the excluded volume effect as one that pushes molecules apart (they cannot interpenetrate). This may be compensated for by an interaction of the sort that makes solute molecules clump together (see Figure 2.3).

The discovery of a θ temperature for a particular macromolecule-solvent system is a prize indeed, for it greatly simplifies all physical studies. The much simpler rules for ideal solutions can then be used.

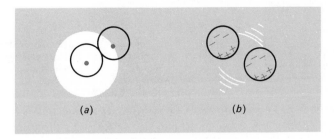

(a)   (b)

Figure 2.3
Excluded volume (a) and intermolecular attraction (b) as determinants of solution behavior.

2.3   MEMBRANE EQUILIBRIA

Semipermeable membranes play a number of important roles in biochemistry. On the practical side, we employ them in dialysis and membrane filtration for the purification of macromolecular substances. Again, equilibrium across membranes is used to measure the binding of small molecules and ions to large molecules (see the subsection on dialysis equilibrium below) or to determine molecular weights. In the cell, membranes take on a similar role, serving to partition regions of the cell, and the cell from its surroundings, by barriers that retain some substances and allow others to pass. The principal difference between the membranes that we employ in the laboratory and those found in the cell (aside from composition) is that the former are *passive*; that is, they act only as barriers and play no active role in the transport of materials. The *active transport* in some cell membranes is a very different thing. Here, at a free-energy price, materials are selectively transported against concentration differences. More will be said of this later.

In this chapter we shall mainly be concerned with the equilibrium phenomena arising from the existence of passive, semipermeable membranes in a system. We shall first write some general rules, then discuss osmotic pressure and the use of dialysis equilibrium, and then turn to the complications introduced when some of the solutes in the solutions carry an electrical charge. At the close, a little will be said about active transport.

Equilibrium Across a Membrane

Consider a solution that is separated from solvent by a semipermeable membrane, as in Figure 2.4. We shall assume that solvent molecules and some low-molecular-weight solute components can pass through the membrane, whereas some kinds of large solute molecules cannot. In practice, the cellophane membranes used most commonly in biochemical laboratories pass materials of $M < 10,000$ and retain larger molecules, but membranes can be prepared that select at much higher or much lower molecular weight.

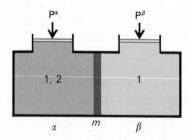

Figure 2.4

Two phases, α and β, are separated by a semipermeable membrane m. The membrane is permeable to component 1 but not to component 2. We can adjust the pressure on α and β as we like.

The system must be regarded as consisting of two phases (α and β), for the fact that the large molecules are restricted to one side means that the system cannot be homogeneous throughout. What are the requirements for equilibrium? We can write them immediately from Equation (2.18):

$$\mu_i^\alpha = \mu_i^\beta \tag{2.46}$$

which holds for all substances that can pass through the membrane. For those substances that cannot pass through, we can make no corresponding statement. Equation (2.46) is the starting point for all discussions of membrane equilibria. For a system composed, for example, of hemoglobin, oxygen, buffer salts, and water, Equation (2.46) would apply to each of the components except hemo-globin.

Osmotic Pressure

Let us consider the implications of Equation (2.46) in a simple system in-volving only a solvent (component 1) that can pass the membrane and a solute (component 2) that cannot. On one side of the membrane (α) we have both 1 and 2; on the other side (β) we have only the solvent, component 1. Both sides are initially assumed to be at the same T and P. The only equation we have is the statement that

$$\mu_1^\alpha = \mu_1^\beta \tag{2.47}$$

at equilibrium. However, this seems to lead to a strange conclusion. According to Equation (2.41), the chemical potential of solvent in a solution will be less than that in pure solvent ($\mu_1 - \mu_1^0 < 0$). It would appear then, that equilibrium will not be attainable, for no matter how much solvent is transferred into phase α, the inequality will still exist. This is actually the case in the system as we have set it up; it is not generally possible for a solution to be in equilibrium with pure solvent at the same T and P. The entropy can always be increased (and the free energy decreased) by transferring some more solvent to further dilute the solution.

To attain equilibrium, we must evidently produce some *other* difference on the two sides of the membrane, so as to produce a chemical potential difference that will compensate for the difference produced by the presence of solute on one side. We might imagine, for example, having the two sides at different temperature. However, it seems impossible to conceive of a membrane that would transmit matter but not heat.† This leaves us only pressure to work with.

Now it is clear that the chemical potential, like any free-energy quantity, will depend on the pressure exerted on a system. Since a membrane can support a pressure difference, suppose we try this. An apparatus we might use is shown in Figure 2.4. We shall keep the pressure on the solvent side (β) at some nominal value, P_0 (1 atm, for example) and increase the pressure on side α (the solution side) to $P_0 + \pi$. What effect will this have on μ_1^α? We know that the variation of free energy with pressure is given by

$$\left(\frac{\partial G}{\partial P}\right)_T = V \tag{2.48}$$

which can be obtained directly from Equation (1.48). We can write for the

---

† If we are willing to accept conditions that only approach equilibrium, this can be done. Thus one type of "osmometer" makes use of the temperature difference established between two droplets, one containing only solvent and the other solution.

partial molar free energy (μ_1) the analogous result

$$\left(\frac{\partial \mu_1}{\partial P}\right)_T = \bar{V}_1 \tag{2.49}$$

where $\bar{V}_1$ is the partial molar volume of solvent. Then we have

$$\mu_1^\beta = \mu_1^0 \tag{2.50}$$

(since on side β we have pure solvent at standard conditions) and

$$\mu_1^\alpha = \mu_1^0 - RTV_1^0 \left\{\frac{C_2}{M_2} + BC_2^2 + \cdots\right\} + \int_{P_0}^{P_0 + \pi} \bar{V}_1 \, dP \tag{2.51}$$

Assuming that $\bar{V}_1 = V_1^0 = \text{constant} = $ the molar volume of pure solvent at 1 atm (a good approximation for low pressures and dilute solutions), we obtain, from (2.51) and (2.50),

$$\mu_1^0 = \mu_1^0 - RTV_1^0 \left\{\frac{C_2}{M_2} + BC_2^2 + \cdots\right\} + V_1^0 \pi$$

or

$$\pi = RT \left\{\frac{C_2}{M_2} + BC_2^2 + \cdots\right\} \tag{2.52}$$

We have calculated the pressure difference required to equate the chemical potential of solvent on the two sides of the membrane. This we call the *osmotic pressure*. It must be applied for the system to be at equilibrium. If the solution is very dilute or if $B = 0$, we obtain

$$\pi = \frac{RTC_2}{M_2} \tag{2.53}$$

This equation shows why π is important; it allows us to measure the molecular weight. Since we shall have to use solutions of finite concentration to measure π and since these may be nonideal, a practical equation for the calculation of M_2 can be written as

$$\frac{\pi}{C_2} = \frac{RT}{M_2} + BRTC_2 + \cdots \tag{2.54}$$

Graphs of osmotic pressure data for globular and unfolded proteins are shown in Figure 2.5. It is evident that extrapolation to $C_2 = 0$ is necessary for accurate results. Some representative data are given in Table 2.2.

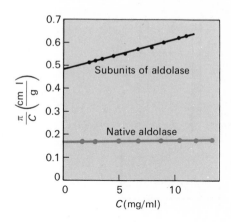

Figure 2.5

Osmotic pressure data for the protein aldolase in buffer at neutral pH and the subunits of this protein in $6\,M$ guanidine hydrochloride. Molecular weights are given in Table 2.2. Note the larger virial coefficient for the unfolded subunits. Data from F. J. Castellino and O. R. Barker, *Biochemistry*, **7**, 2207 (1968). Reprinted by permission of the American Chemical Society.

TABLE 2.2 **SOME MOLECULAR WEIGHTS OBTAINED FROM OSMOTIC PRESSURE STUDIES**

Substance	M_n
Ovalbumin	44,600
Hemoglobin	66,500
Aldolase	156,500
Aldolase subunits	42,400
Amylose, various samples	32,000–150,000

One can measure the osmotic pressure in a number of ways. For example, each side of the apparatus depicted in Figure 2.4 could be fitted with a standpipe, and solvent could be allowed to flow from β to α until the hydrostatic pressure difference equaled π [the static osmometer, Figure 2.6(*a*)]. Alternatively, we could adjust the pressure difference until no flow of solvent occurred [the dynamic osmometer, Figure 2.6(*b*)]. The latter method is used in modern, automated osmometers.

Equation (2.54) shows how the molecular weight of a homogeneous solute can be measured. Now suppose the solute is heterogeneous, that is, that it consists of a mixture of macromolecular components. What result is obtained then? At very low concentrations the osmotic pressure can be regarded as the sum of individual contributions:

$$\pi = \sum_i \pi_i \tag{2.55}$$

If

$$\pi_i = RTC_i/M_i \tag{2.56}$$

we obtain

$$\pi = RT\sum_i C_i/M_i \tag{2.57}$$

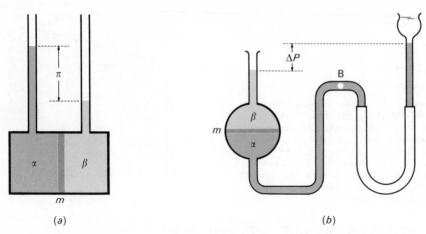

Figure 2.6   Osmometers. (*a*) The static type, in which solvent transfer occurs until a hydrostatic pressure difference equal to π is developed. (*b*) The dynamic type, in which the pressure difference is adjusted until the bubble, B, is stationary, showing that there is no solvent transfer occurring. This can be automated by a servomechanism.

This does not look very useful. But suppose we divide and multiply the right by $\sum\limits_{i} C_i = C$, the total solute concentration. Then

$$\pi = RTC \left/ \left[\sum_{i} C_i \middle/ \sum_{i} (C_i/M_i) \right] \right.$$

(2.58)

This looks like Equation (2.53). If we define an *average molecular weight*, M_n, by

$$M_n = \sum_{i} C_i \middle/ \sum_{i} (C_i/M_i)$$

(2.59)

we have

$$\pi = \frac{RTC}{M_n}$$

(2.60)

or, for finite concentrations,

$$\pi = RT \left\{ \frac{C}{M_n} + \bar{B}C^2 + \cdots \right\}$$

(2.61)

where $\bar{B}$ is an average virial coefficient. These important results tell us that the osmotic pressure measures an average of the molecular weights in a heterogeneous solute and how that average is defined. We shall encounter average molecular weights in many situations. A number of useful averages are tabulated in Table 2.3, expressed in terms of both weight and number concentrations. Note that the average measured by osmotic pressure is just the mean molecular weight, on a number concentration basis. This average will be obtained from any colligative property (such as osmotic pressure, freezing-point depression, or boiling-point elevation) that depends on the number of solute molecules per unit volume.

TABLE 2.3   AVERAGE MOLECULAR WEIGHTS

	Definition[a]	
Average	In terms of N_i	In terms of C_i
Number average, M_n	$\sum_i N_i M_i / \sum_i N_i$	$\sum_i C_i / \sum_i (C_i/M_i)$
Weight average, M_w	$\sum_i N_i M_i^2 / \sum_i N_i M_i$	$\sum_i C_i M_i / \sum_i C_i$
Z average, M_z	$\sum_i N_i M_i^3 / \sum_i N_i M_i^2$	$\sum_i C_i M_i^2 / \sum_i C_i M_i$

[a] C_i is the concentration in weight per volume; N_i is the concentration in numbers of moles or molecules per unit volume.

Dialysis Equilibrium

Frequently, use is made of the membrane equilibrium to measure the binding of small molecules or ions to a macromolecule. The principle is easily visualized. The macromolecular solution is placed inside a membrane bag (phase α) and suspended in a solution containing the small molecules (phase β) (Figure 2.7). At equilibrium, any excess in concentration of the low-molecular-weight substance on the macromolecular side of the membrane is taken as evidence for binding. While this appears straightforward, a closer analysis is easy. Calling the low-molecular-weight substance (which we presume to be a nonelectrolyte) component 2, we have at equilibrium

$$\mu_2^\alpha = \mu_2^\beta \tag{2.62}$$

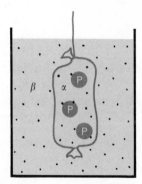

Figure 2.7
Dialysis equilibrium. The small dots represent solute molecules that can pass through the membrane. If some are bound by the protein P, the total inside concentration of these molecules will be greater than that outside.

Neglecting the very small difference in chemical potential resulting from the equilibrium pressure difference between the two sides, we obtain, from Equation (2.33),

$$RT \ln C_2^\alpha y_2^\alpha = RT \ln C_2^\beta y_2^\beta \tag{2.63}$$

or

$$C_2^\alpha y_2^\alpha = C_2^\beta y_2^\beta \tag{2.64}$$

Note that to this point we have made no distinction between bound and unbound component 2 on side α. However, if we assume that there exists, on side α, both bound and unbound component 2, and that the environments are sufficiently similar that the activity coefficient for free component 2 is the same on both sides, we have

$$C_2^\alpha(\text{unbound}) = C_2^\beta \tag{2.65}$$

and thus

$$C_2^\alpha(\text{bound}) = C_2^\alpha - C_2^\beta \tag{2.66}$$

It is clear that a number of somewhat arbitrary assumptions have gone into this result. It is not surprising, therefore, that the numbers obtained from such analysis can even be such as to indicate negative binding. In fact, a simple mechanical picture leads to the expectation of such results. Suppose that the macromolecule does not bind the solute at all but simply excludes it from the volume occupied by the macromolecules. Then $C_2^\alpha < C_2^\beta$, and we would interpret this as negative binding. It is clear that unless binding is very strong, analyses depending on the above assumptions must be treated with caution.

Effects of Solute Charge: The Donnan Equilibrium

So far, we have assumed that the substances involved in membrane equilibria are uncharged. If this is not the case and the macromolecules are polyelectrolytes, the situation becomes more complicated. Since polyelectrolyte behavior is the rule rather than the exception with biopolymers, such complications are by no means trivial. All proteins and nucleic acids, for example, are polyelectrolytes. In this section we shall consider only the behavior of strong polyelectrolytes and for concreteness will take the case of a polymer that dissociates according to the rule

$$\text{PX}_z \longrightarrow \text{P}^{+z} + z\text{X}^- \tag{2.67}$$

where X^- is a monovalent anion. The general result is symmetric to charge type, so we do not lose generality by taking the macromolecular species P^{+z} to be a polycation.

If the polycation and its counterions are the only electrolytes present in the system, the result is trivial. Since the counterions must remain on the same side of the membrane as the polymer (to maintain electrical neutrality), we simply have $z + 1$ particles present for every solute molecule introduced. All colligative properties, including osmotic pressure, should then be increased by a factor of $z + 1$, and the material will behave as if it has a molecular weight $1/(z + 1)$ of the undissociated polymer. We have taken the number-average molecular weight of the mixture of polycations and counterions.†

Fortunately, this rather disastrous situation is rarely approached, for polyelectrolytes are almost always studied in the presence of some additional low-molecular-weight electrolyte, and this markedly changes the results. Let us consider, then, the more complex system shown in Figure 2.8, in which a $1:1$ electrolyte $(BX \longrightarrow B^+ + X^-)$ has been added to the system. The membrane is assumed to be permeable to B^+, X^-, and water but impermeable to P^{+z}. If this is true, we must have

$$\mu_{BX}^{\alpha} = \mu_{BX}^{\beta} \tag{2.68}$$

We shall assume the solution to be ideal. The chemical potential of the salt can then be written as $\mu_{BX} = \mu_{BX}^0 + RT \ln c_B c_X$, where c_B and c_X are the molar concentrations of B and X. Then, neglecting the difference $\mu_{BX}^{0\alpha} - \mu_{BX}^{0\beta}$, we have, from (2.68),

$$c_B^{\alpha} c_X^{\alpha} = c_B^{\beta} c_X^{\beta} \tag{2.69}$$

A second equation is obtained from the requirement for electrical neutrality:

$$
\begin{array}{cc}
\text{On Side } \alpha & \text{On Side } \beta \\
zc_P^{\alpha} + c_B^{\alpha} = c_X^{\alpha} & c_B^{\beta} = c_X^{\beta} \\
+ \quad\quad - \quad\quad + \quad\quad - &
\end{array}
\tag{2.70}
$$

where c_P is the molar concentration of polymer.

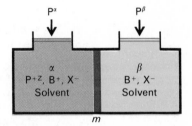

Figure 2.8
Membrane equilibrium with a polyelectrolyte solute. See the text.

† In aqueous solutions the result is actually a bit more complicated, for dissociation of water must also be considered.

Combining (2.69) and (2.70), we obtain

$$(c_B^\beta)^2 = c_B^\alpha(c_B^\alpha + zc_P^\alpha) \tag{2.71}$$

$$(c_X^\beta)^2 = c_X^\alpha(c_X^\alpha - zc_P^\alpha) \tag{2.72}$$

or

$$c_B^\alpha - c_B^\beta = \frac{-zc_P^\alpha c_B^\alpha}{(c_B^\alpha + c_B^\beta)} \tag{2.73}$$

$$c_X^\alpha - c_X^\beta = \frac{zc_P^\alpha c_X^\alpha}{(c_X^\alpha + c_X^\beta)} \tag{2.74}$$

Thus at equilibrium the counterion (X) will be more concentrated on the polymer side of the membrane, while the cation B will be less. The salt has been redistributed so as to keep electrical neutrality. The concentration differences given by (2.73) and (2.74) will contribute to the osmotic pressure. Remembering that we have assumed ideal behavior and that we are using molar concentrations, π will be written as

$$\pi = RT\{c_P^\alpha + (c_B^\alpha - c_B^\beta) + (c_X^\alpha - c_X^\beta)\} \tag{2.75}$$

After insertion of Equations (2.73) and (2.74) and rearrangement, this gives

$$\pi = RT\{c_P^\alpha + z^2(c_P^\alpha)^2/(c_B^\alpha + c_B^\beta + c_X^\alpha + c_X^\beta)\} \tag{2.76}$$

In the "correction term" to the right in Equation (2.76) we can usually make the assumption that $c_B^\alpha \simeq c_B^\beta \simeq c_X^\alpha \simeq c_X^\beta \simeq c_{BX}$, where c_{BX} is the initial (molar) concentration of the salt BX in the system. Then

$$\pi \simeq RT\{c_P^\alpha + z^2(c_P^\alpha)^2/4c_{BX}\} \tag{2.77}$$

or, if we write $c_P^\alpha = C_P/M_P$, where C_P is a weight concentration,

$$\pi = RT\left\{\frac{C_P}{M_P} + \frac{z^2 C_P^2}{4M_P^2 c_{BX}}\right\} \tag{2.78}$$

In other words, the solution acts as if it were nonideal, with a virial coefficient proportional to the square of the charge/mass ratio. This effect can be large for strong polyelectrolyte or for proteins far from the isoelectric point.

A second consequence of the Donnan equilibrium is seen in Equations (2.73) and (2.74). If we were attempting to measure binding of ions by a protein and neglected the Donnan effect, we would reach the conclusion (in this case) that X^- ions were bound and B^+ ions excluded. This means that all ion-binding experiments must be corrected for this membrane effect.

Finally, one should point out the usefulness of high electrolyte concentration. It is evident from Equation (2.78) that the osmotic pressure will approach the ideal value as c_{BX} is made larger. Since analogs of the Donnan effect exist in such other physical measurements as light scattering and sedimentation, the advisability of using moderately high salt concentrations in any study of a charged polymer should be obvious. In some sedimentation studies of serum albumin, which has a molecular weight of 68,000, experiments at low salt and low pH (high charge) gave apparent molecular weights as low as 3,000.

2.4   ACTIVE TRANSPORT

From what has been said in the preceding sections, it would seem incredible that a membrane could spontaneously and selectively transport molecules or ions so as to *produce* large concentration differences. Yet this is so, and even the example used in Chapter 1, the diffusion of sugar out of a dialysis bag, can be neatly reversed by certain bacteria that can spontaneously transport sugar inward through their cell walls even though the outside concentration is much less than that within the bacteria. How can this be so? Does the second law not always hold? For surely we have shown that the free-energy change accompanying the transport of material against a concentration gradient is positive.

This is, of course, true, but when such processes are observed, we must consider other, more subtle possibilities rather than abandon the second law. A first possibility is suggested by the earlier discussion of dialysis equilibrium. It is actually a gradient in chemical potential that determines the spontaneous direction of free diffusion, and we may be observing a situation in which most of the solute on the high-concentration side of the membrane is actually complexed or bound in some way. If this fact is not known (and it is not always easy to detect such binding in a complicated mixture), we may think that the concentration of free solute is much higher than it actually is; the free solute may actually be more dilute than on the "low-concentration" side of the membrane.† Such situations are by no means rare; the scavenging of trace metals from sea water by organisms often operates by such mechanisms.

Such behavior is not true active transport. There remains a class of phenomena in which an actual chemical potential difference is established and maintained. But even these involve no violation of the laws of thermodynamics. The clue is found in the fact that actively metabolizing membranes are required. Evidently, the process of transport across a membrane (which costs a free-energy price when the chemical potential gradient opposes it) can be coupled to a spontaneous reaction that has a larger negative free-energy change.

---

† To take an example, the Mg^{2+} concentration in most cells is much greater than in the surrounding fluids, yet this does not prove active transport, for most Mg^{2+} are strongly bound.

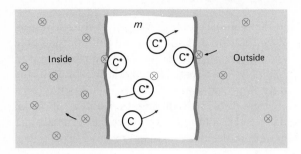

Figure 2.9
A simple model for active transport. See the text.

Suppose we write the transport of 1 mole of substance X as

$$X(\text{at } \mu_X) \longrightarrow X(\text{at } \mu'_X) \tag{2.79}$$

$$\Delta G_{tr} = \mu'_X - \mu_X \tag{2.80}$$

If we consider a reactive carrier that can react with X on one side of the membrane (see Figure 2.9)

$$C^* + X \longrightarrow C^*X \tag{2.81}$$

$$\Delta G_2 = \mu_{C^*X} - \mu_{C^*} - \mu_X \tag{2.82}$$

and then release X by a catalyzed reaction on the other side of the membrane

$$C^*X \longrightarrow C + X \tag{2.83}$$

$$\Delta G_3 = \mu'_X + \mu_C - \mu_{C^*X} \tag{2.84}$$

the overall change becomes

$$\Delta G = \Delta G_3 + \Delta G_2 \tag{2.85}$$

$$= \mu'_X + \mu_C - \mu_{C^*X} + \mu_{C^*X} - \mu_{C^*} - \mu_X$$

$$= (\mu'_X - \mu_X) + (\mu_C - \mu_{C^*}) \tag{2.86}$$

so even if $\mu'_X - \mu_X > 0$, ΔG can be negative if $\mu_C - \mu_{C^*}$ is very negative, that is, if the transition from C* to C is a spontaneous process.

Such a scheme is just another example of the kind of "coupled reaction" so common in biochemistry. A highly favored reaction (the conversion of C* to C) is used to drive the unfavored process (the active transport of X). The free-energy change for the overall process is negative, which is all that thermodynamics asks. Of course, the formation of the unstable *activated carrier* has itself required the expenditure of free energy, and thus when the membrane is not actively metabolizing, the active transport must stop, and the whole system reverts to equilibrium with $\mu'_X = \mu_X$.

Viewed in this way, active transport is not paradoxical or even surprising. All of the anabolic processes in life, the organizing, segregating, and arranging that goes on, are, taken one by one, nonspontaneous processes. The assembly of a protein molecule from a collection of amino acids involves an enormous free-energy *increase*. But all of these processes are linked and depend ultimately on the utilization of a little of the energy of sunlight to drive them, for a while, in nonspontaneous directions. But the entropy price is always paid on the grand scale. For every algal cell that lives, a little more of the sun's energy is absorbed, the earth heats a bit more, and the energy of the universe is a little more evenly distributed.

PROBLEMS

1.* Using the Gibbs-Duhem equation, obtain an expression for $\mu_2 - \mu_2^0$ from the equation

$$\mu_1 - \mu_1^0 = -RTV_1^0 C_2\left(\frac{1}{M_2} + BC_2\right)$$

2. Bushy stunt virus particles are approximately spherical and of radius 140 Å, and $M = 1.07 \times 10^7$. Calculate, on the basis of excluded volume theory, the second virial coefficient of bushy stunt virus.

3.* Derive an expression for ΔH_{mix} for a polypeptide with dichloroacetic acid (DCA). Assume that DCA in the liquid state is entirely H-bonded according to the structure

$$HCCl_2 - C \overset{\displaystyle O\cdots HO}{\underset{\displaystyle OH\cdots O}{\Big\langle}} C - CCl_2H$$

and that the polypeptide in the solid state is entirely in the β configuration. Denote the enthalpy contribution from various H-bond types by

=O···HO ε_{OHO}

=O···HN ε_{OHN}

Consider H bonding to be the *only* type of interaction and assume that each amino acid residue in the peptide can be considered as a "segment," equal in volume to a DCA molecule. Under what conditions will the mixing process be exothermic? [*Hint*: See pp. 207 and 208, Tanford (1961).]

4. The following data describe the osmotic pressure (in centimeters of solution) of solutions of a fraction of polyvinyl chloride in methyl ethyl ketone (MEK) at 25°C. The density of MEK is 0.80 g/ml, and the solution may be assumed to be of equal density.

Calculate $\overline{M}_n$ and the second virial coefficient B for the sample.

$C(g/100\ cm^3)$	$\pi\ (cm\ solution)$
0.167	0.40
0.489	1.44
0.654	2.10
0.801	2.76
0.980	3.83

5. Obtain an equation for the freezing-point depression of a polymer solution in terms of M and B and estimate the freezing-point depression of a solution of 1 g/100 ml of the bushy stunt virus sample described in Problem 2.

6. Estimate the second virial coefficient (B) for DNA in 0.5 M NaCl at a pH where all phosphate residues are charged. Take $v_2 = 0.53$ and assume the DNA to be double-stranded, with a molecular weight of 3×10^6 for each strand. Estimate the ratio of π/C at 1.0 g/100 ml to the value at $C = 0$. All data at 25°C. You may assume a weight/length ratio of 149 molecular-weight units/Å and consider DNA a rigid rod. Assume that contributions to B are additive, including the Donnan term.

7. Calculate the free-energy change involved in transporting 1 mole of sucrose from a region where the concentration is 0.001 M to a region where it is 1 M.

REFERENCES

An excellent, detailed discussion of solution thermodynamics in general and solutions of macromolecules in particular is found in:

Tanford, C.: *Physical Chemistry of Macromolecules*, John Wiley & Sons, Inc., New York, 1961.

The fundamentals of the thermodynamics of polymer solutions are detailed in:

Flory, P.: *Principles of Polymer Chemistry*, Cornell University Press, Ithaca, N.Y., 1953.

THREE | CHEMICAL EQUILIBRIUM

Biochemistry is concerned with many types of chemical equilibria. Some of these involve the small molecules metabolized and utilized as monomers in construction of the cell. Other important equilibria involve macromolecules, either in reactions with other macromolecules or in reactions with small molecules. All of these reactions occur in solution, and the usual approach of physical chemistry texts, to emphasize gas-phase reactions, is of little direct interest to the biochemist. In this chapter we shall discuss solution equilibria in general and then treat in detail the general class of multiple equilibria involving macromolecules and small molecules. A brief discussion of order-disorder equilibria in macromolecules will complete the chapter. Further details of the chemical equilibria of metabolism are given in Larner, in this series, and a number of important cases of macromolecule-macromolecule equilibria are described in Wold, this series.

3.1 THERMODYNAMICS OF CHEMICAL REACTIONS IN SOLUTION

As emphasized in Chapter 2, discussion of the thermodynamics of solutions centers on the chemical potential. Suppose we have a general chemical reaction

$$aA + bB + \cdots \rightleftharpoons gG + hH + \cdots \qquad (3.1)$$

in which a moles of A, b moles of B, and so forth, at molar concentration c_A, c_B, and so forth, are converted into g moles of G, h moles of H, and so forth, at concentrations c_G, c_H, and so forth.† If we presume that the reaction occurs at constant pressure and temperature, we may write the free-energy change, ΔG, as

$$\Delta G = G(\text{final state}) - G(\text{initial state}) \tag{3.2}$$

or, by Equation (2.10),

$$\Delta G = g\mu_G + h\mu_H + \cdots - a\mu_A - b\mu_B - \cdots \tag{3.3}$$

where μ_G, μ_H, and so forth are the chemical potentials at the concentrations existing in the corresponding solutions. Recalling that

$$\mu_i = \mu_i^0 + RT \ln c_i y_i \tag{3.4}$$

we may rewrite (3.3) as

$$\Delta G = g\mu_G^0 + h\mu_H^0 + \cdots - a\mu_A^0 - b\mu_B^0 - \cdots + RT \ln \frac{(c_G y_G)^g (c_H y_H)^h \cdots}{(c_A y_A)^a (c_B y_B)^b \cdots}$$

$$= (g\mu_G^0 + h\mu_H^0 + \cdots - a\mu_A^0 - b\mu_B^0 - \cdots) + RT \ln \frac{c_G^g c_H^h \cdots}{c_A^a c_B^b \cdots} + RT \ln \frac{y_G^g y_H^h \cdots}{y_A^a y_B^b \cdots} \tag{3.5}$$

The three terms in this equation may be distinguished as follows: The first, involving the μ_i^0's, is the standard-state free-energy change (ΔG^0). It corresponds to the ΔG that would be observed if a moles of A and so forth in the standard state formed g moles of G and so forth, also in the standard state. The second term, involving the c_i's, gives the effect of the actual initial and final concentrations on the free-energy change. It is corrected by the third term, which involves only the activity coefficients. For purposes of discussion, we may assume this activity coefficient term to be zero; this would be true, for example, if all components were present in such great dilution that they could be considered to behave ideally. In this event, all y_i's will equal unity, and the third term will vanish. We may then rewrite (3.5) as

$$\Delta G = \Delta G^0 + RT \ln \frac{c_G^g c_H^h \cdots}{c_A^a c_B^b \cdots} \tag{3.6}$$

The values of ΔG^0 for various reactions have been tabulated (see Table 3.1 for some examples). While such data are useful, it should be emphasized that they do *not* represent the ΔG values accompanying reactions under all possible conditions. In fact, since the standard states are usually defined (at least for

---

† Note that we are not assuming equilibrium here; we simply take reactants under stated conditions and convert them to products.

TABLE 3.1 **STANDARD FREE-ENERGY CHANGES ACCOMPANYING SOME REACTIONS OF BIOCHEMICAL IMPORTANCE**[a]

I. Hydrolysis reactions. Based on a standard state of 1-M concentration of reactants (except H^+, when pH is given). Water activity assumed unity.

Reactant	pH	Other conditions	$-\Delta G^0$ (kcal/mole)
Phosphoenolpyruvate	7.0		14.8
Carbamyl phosphate	9.5		12.3
3-Phosphoglycerol phosphate	6.9		11.8
Creatine phosphate	7.0	37°C	10.3
ATP ($\rightarrow$AMP + PPi)	7.0	Excess Mg^{2+}	7.7
ATP ($\rightarrow$ADP + Pi)	7.0	37°C, excess Mg^{2+}	7.3
Glucose-1-phosphate	7.0	25°C	5.0
Pyrophosphate	7.0	0.005 M Mg^{2+}	4.5
Glucose-6-phosphate	7.0	25°C	3.3
α-Glycerophosphate	8.5	38°C	2.2
Sucrose			7.0
Maltose			4.0
Glycogen			4.0
Glycylglycine		37.5°C	3.6
Benzoyltyrosyl-glycinamide	6.5	23°C	0.42

II. Heats and free energies of formation for formation of compound in standard state from C (graphite), H_2 (gas), O_2 (gas), N_2 (gas), all in standard states.

Substance	$-\Delta H_f^0$	$-\Delta G_f^0$
Glycine	128.4	90.3
L-Leucine	152.4	83.1
L-Tryptophan	99.2	28.5
Glycylglycine	178.1	116.6
DL-Leucylglycine	205.6	112.4
Glycyl-DL-tryptophan	148.8	54.7
Water (l)	68.32	56.69

[a] Most data are from H. Sober (ed.), *The Handbook of Biochemistry*, Chemical Rubber Co., Cleveland, Ohio, 1968.

low-molecular-weight materials) as unit molarity, the conditions for the standard-state reactions are usually ludicrous by biological standards. To make this clear, let us take a common example. The hydrolysis of adenosine triphosphate (ATP) to yield adenosine diphosphate (ADP) and inorganic phosphate (Pi) may be written

$$ATP + H_2O \rightleftharpoons ADP + Pi \tag{3.7}$$

so we might write

$$\Delta G = \Delta G^0 + RT \ln \frac{c_{ADP} c_{Pi}}{c_{ATP}} \tag{3.8}$$

Note that the water concentration has not been included. The solutions are usually so dilute that any water consumed in the reaction represents a negligible change in the total water concentration.† The term $(-RT \ln c_{H_2O})$ has, in fact, been included in the ΔG^0 term, since it is so very nearly a constant. Now ΔG^0 for this reaction at 25°C is, by the best estimate, about -7 kcal at pH 7.5. However, in the cell, where such reactions are of interest to us, the conditions are far from those in the standard state. Concentrations are more likely to be near $10^{-2}M$. In this case, for a reaction in which ATP at $10^{-2}M$ is hydrolyzed to ADP and P at the same concentration, we obtain

$$\Delta G \simeq -10 \text{ kcal/mole} \tag{3.9}$$

What this means, of course, is that ATP hydrolysis is promoted at high dilutions.

In the above, we have been calculating the free-energy change accompanying a given reaction involving given initial and final conditions. Suppose, instead, that we ask that the reaction be carried out reversibly? This means that equilibrium concentrations are maintained. In practice, for a finite amount of reaction to be carried out in this way, it would be necessary to add reactants and take away products so as to maintain the unique set of equilibrium concentrations. In this case, at constant T and P, the total free-energy change for any amount of reaction will be zero. Therefore, Equation (3.6) becomes

$$0 = \Delta G^0 + RT \ln \left(\frac{c_G^g c_H^h \cdots}{c_A^a c_B^b \cdots} \right)_{eq} \tag{3.10}$$

for ideal solutions, and Equation (3.5) yields

$$0 = \Delta G^0 + RT \ln \left[\frac{(c_G y_G)^g (c_H y_H)^h \cdots}{(c_A y_A)^a (c_B y_B)^b \cdots} \right]_{eq} \tag{3.11}$$

for the more general case. Since the concentrations involved must now be the equilibrium concentrations, we have the result

$$\Delta G^0 = -RT \ln \left[\frac{(c_G y_G)^g (c_H y_H)^h \cdots}{(c_A y_A)^a (c_B y_B)^b \cdots} \right]_{eq} = -RT \ln K_t \tag{3.12}$$

This means that the ΔG^0 value will tell us the equilibrium constant for a reaction. This will, in general, be the thermodynamic equilibrium constant, which involves the activity coefficient correction. It can, for dilute solutions, be approximated by the practical equilibrium constant, defined as

$$K = \left(\frac{c_G^g c_H^h \cdots}{c_A^a c_B^b \cdots} \right)_{eq} \tag{3.13}$$

---

† Is this always true in vivo? It is usually assumed that the in-vitro conditions used to study biochemical reactions correspond roughly to those in the cell. But we really know little about such questions as the concentrations of reactants (including water!) in compartmentalized regions of the cell.

Students are frequently confused at this point. Why is it that the *standard-state* free-energy change can be used to measure the *equilibrium* constant, while the free-energy change at equilibrium is zero? It helps to visualize Equation (3.10) or (3.11) as stating something like this: We can pick an arbitrary reference state, the standard state. In general, ΔG will not be zero if reactants and products are in standard states.

There will be some set of concentrations (the equilibrium set) at which ΔG will be zero. The second term in (3.10) or (3.11) represents the free-energy change involved in bringing reactants and products from the standard-state concentrations to the equilibrium-state concentrations. This term must just balance ΔG^0 for the overall ΔG to be zero.

For the purpose of interpreting values, it is worthwhile to see how the equilibrium constant varies with ΔG^0: We can rewrite Equation (3.12) as

$$K_t = e^{-\Delta G^0/RT} \tag{3.14}$$

Since $RT \simeq 0.6$ kcal near room temperature, a value of $\Delta G^0 = -0.6$ kcal corresponds to $e^1 = 2.718$, and a value of $\Delta G^0 = -1.2$ kcal corresponds to $K = e^2$, about 8.7. Thus even what seem to be small ΔG^0 values correspond to fairly large values of the equilibrium constant. If we consider an "essentially irreversible" reaction to be one in which $K > 10^4$, this will correspond to a free-energy change more negative than about 5.5 kcal. In other words, such reactions as the hydrolysis of ATP are essentially irreversible.

Another way of measuring the free-energy change in a chemical reaction is through the electrical potential of a cell (real or hypothetical) in which the reaction occurs. It is always possible to calculate the electromagnetic force corresponding to a given reaction, since the free-energy change determines the maximum work other than PV work that the reaction can produce (see Figure 1.4). If we interpret this "other" work as electrical, we have

$$\Delta G = -w_{\text{rev}} = -n\mathscr{F}\mathscr{E} \tag{3.15}$$

where n is the number of moles of electrons transferred per mole of reaction, and $\mathscr{F}$ is the Faraday constant (23.06 kcal/V. equiv). This leads to the Nernst equation

$$\mathscr{E} = \mathscr{E}^0 - \frac{RT}{n\mathscr{F}} \ln \frac{(c_G y_G)^g (c_H y_H)^h \cdots}{(c_A y_A)^a (c_B y_B)^b \cdots} \tag{3.16}$$

where $\mathscr{E}^0$ is the standard-state potential. A table of standard potentials is equivalent to a table of standard free-energy charges, since

$$\Delta G^0 = -n\mathscr{F}\mathscr{E}^0 \tag{3.17}$$

and a positive value for the standard potential for a process indicates spontaneity under standard conditions. Biological oxidation-reduction reactions are often thought of in this way.

For many reasons, it is often important to be able to determine the thermo-dynamic parameters (ΔG^0, ΔH^0, ΔS^0) involved in biochemical processes. Thus a few words about the experimental problems are in order. If it is possible to experimentally determine the equilibrium constant for a reaction, by measuring the concentrations of reactants and products present at equilibrium, Equation (3.12) yields ΔG^0 directly. Two points should be noted: First, in most bio-chemical systems, activity coefficients are not known; thus K_t in Equation (3.12) must be approximated by the practical equilibrium constant [Equation (3.13)]. Second, the meaning of the value of ΔG^0 obtained will depend on the way in which the reaction is written and the way in which the equilibrium constant is consequently expressed. To take an example: If we express a dimerization reaction by the equation

$$2A \rightleftharpoons B \tag{3.18}$$

and consequently write

$$K = \frac{c_B}{c_A^2} \tag{3.19}$$

the ΔG^0 obtained will be for the process as written in (3.18): Two moles of A (in standard state) forming 1 mole of B. On the other hand, we could equally well write

$$A \rightleftharpoons \tfrac{1}{2}B \tag{3.20}$$

and

$$K' = \frac{c_B^{1/2}}{c_A} \tag{3.21}$$

Then the ΔG^0 calculated from K' will be the free-energy change corresponding to the conversion of 1 mole of A into $\frac{1}{2}$ mole of B. It will, of course, be half as large a number as given from (3.19). This is all eminently sensible, but it some-times causes confusion.

To obtain the enthalpy change corresponding to a given reaction, two courses are available. The most direct way is simply to let the reaction occur in a calorimeter and measure the heat evolved or absorbed. One must, of course, know the number of moles that actually react, and the reaction must usually be fairly rapid to allow precise measurement; it is difficult to measure a small amount of heat that is only slowly evolved. To be most precise, of course, one should correct to the standard states of reactants and products (to obtain ΔH^0) by taking into account the heats of dilution of products and reactants. In practice, such corrections are often neglected.

For reactions that are simply not amenable to calorimetric study, more indirect ways of determining ΔH^0 are available. For example, Equation (3.12) can be rewritten as

$$-RT \ln K = \Delta G^0 = \Delta H^0 - T \Delta S^0 \tag{3.22}$$

or

$$\ln K = \frac{-\Delta H^0}{RT} + \frac{\Delta S^0}{R} \tag{3.23}$$

(Note that we have made the practical approximation that $K_t \simeq K$; this will be used henceforth.) Equation (3.23) states that $\ln K$ should be a linear function of $1/T$ *if* ΔH^0 and ΔS^0 are independent of T. Many times this assumption is valid; often it is not but is made anyway, with unfortunate results. The slope of the straight line obtained by graphing $\ln K$ versus $1/T$ (called a van't Hoff graph) should be $-\Delta H^0/R$.

It must be emphasized that this technique, as it is often employed, is beset with traps. Often ΔH^0 varies with T; this will happen, for example, whenever there is a difference in heat capacity between reactants and products. The occurrence of the *logarithm* of K in Equation (3.23) means that any physical quantities *proportional* to the concentrations of reactants and products can be used to define an apparent K, which will still give the right slope. But a hazard exists if these quantities do not measure what the experimeter thinks they measure or if the reaction is a complex one. One may literally measure a "standard-state free-energy change" for a completely meaningless reaction.

3.2   MULTIPLE EQUILIBRIA

A macromolecule, being large and complex, may have a number of sites for interaction with small molecules. Many such interactions involve reversible binding of small *ligands* to the macromolecule. Such *multiple equilibria* range over a broad spectrum; examples include the binding of oxygen by hemoglobin, the binding of protons by all proteins, and the binding of magnesium ions by DNA. In some cases the binding sites act quite independently of one another. Such binding is relatively uninteresting, for, as we shall see, it is indistinguishable from the interactions of low-molecular-weight substances. But in a large number of important cases the binding is cooperative, the occupation of one site influencing the strength of binding at other sites. In a teleological sense, the existence of cooperative processes may be regarded as one reason for cells to have macromolecules.† Such processes are not found with low-molecular-weight substances in solution.

---

† Cooperativity is useful, for it allows (1) control of processes and (2) an on-off character to some processes. Compare the response of myoglobin and hemoglobin to changes in oxygen pressure in Figure 3.7.

Let us consider first a very general case. A macromolecule, P, which is assumed to have n binding sites, reacts with a low-molecular-weight substance, A. We may write the reactions and the corresponding association and dissociation constants as follows:

Reactions	Association Constant	Dissociation Constant
$P + A \rightleftharpoons PA$	$K_1 = \dfrac{[PA]}{[P][A]}$	$\mathbf{K}_n = \dfrac{[P][A]}{[PA]}$
$PA + A \rightleftharpoons PA_2$	$K_2 = \dfrac{[PA_2]}{[PA][A]}$	$\mathbf{K}_{n-1} = \dfrac{[PA][A]}{[PA_2]}$
$\vdots$	$\vdots$	$\vdots$
$PA_{n-2} + A \rightleftharpoons PA_{n-1}$	$K_{n-1} = \dfrac{[PA_{n-1}]}{[PA_{n-2}][A]}$	$\mathbf{K}_2 = \dfrac{[PA_{n-2}][A]}{[PA_{n-1}]}$
$PA_{n-1} + A \rightleftharpoons PA_n$	$K_n = \dfrac{[PA_n]}{[PA_{n-1}][A]}$	$\mathbf{K}_1 = \dfrac{[PA_{n-1}][A]}{[PA_n]}$

$$(3.24)$$

The square brackets denote molar concentrations. It should be noted that *either* the set of dissociation or association constants may be used to describe the reactions. Both are given because some kinds of experiments are traditionally described in terms of one set and other kinds in the other. Also, it should be mentioned that these are not thermodynamic equilibrium constants (which would involve activity coefficients) but practical constants. With the complexities and uncertainties involved in studies of macromolecules, it is common to overlook the difference.

In most cases it is not possible to determine the concentrations of individual species in such an equilibrium mixture. Rather, most experimental methods yield only the number of moles of A that are bound per macromolecule. Since the population will be heterogeneous, this is an average quantity

$$\bar{v} = \frac{[A]_{bound}}{[P]_{total}} \tag{3.25}$$

where $[P]_{total}$ denotes the total number of moles of macromolecule in all forms (P, PA, PA$_2$, and so forth). Since total number of sites is $n[P]_{total}$, we may also define the fraction of sites occupied by

$$\theta = \frac{[A]_{bound}}{n[P]_{total}} \tag{3.26}$$

or

$$\theta = \frac{\bar{v}}{n} \tag{3.27}$$

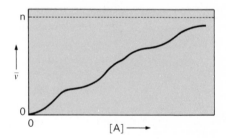

Figure 3.1
A schematic graph of $\bar{v}$ versus [A] in some complicated binding process. Two points are general: $\bar{v} \to 0$ as [A] $\to 0$, and $\bar{v} \to n$ as [A] $\to \infty$.

Finally, in some cases it is useful to consider an average dissociation number, $\bar{r}$,

$$\bar{r} = n - \bar{v} \tag{3.28}$$

which is obviously the average number of sites per molecule that are *not* binding A.

It is obvious that a graph of $\bar{v}$ versus [A] should approach zero at [A] $\longrightarrow 0$ and approach n as [A] $\longrightarrow \infty$ (Figure 3.1). However, the exact form will depend on the nature of binding. This we shall now examine. With the aid of Equation (3.24), a general formulation for $\bar{v}$ and $\bar{r}$ is possible. We note that

$$[A]_{bound} = [PA] + 2[PA_2] + 3[PA_3] + \cdots \tag{3.29}$$

since PA_2 binds two molecules of A and so forth and that

$$[P]_{total} = [P] + [PA] + [PA_2] + \cdots \tag{3.30}$$

Furthermore, since

$$[PA] = K_1[P][A], \qquad [PA_2] = K_1 K_2[P][A]^2, \cdots$$

we can obtain the general expression for $\bar{v}$

$$\bar{v} = \frac{K_1[A] + 2K_1K_2[A]^2 + \cdots + n(K_1K_2\cdots K_n)[A]^n}{1 + K_1[A] + K_1K_2[A]^2 + \cdots + (K_1K_2\cdots K_n)[A]^n} \tag{3.31}$$

which is sometimes called the Adair equation.

Similarly (as the reader should convince himself),

$$\bar{r} = \frac{K_1/[A] + 2K_1K_2/[A]^2 + \cdots + n(K_1K_2\cdots K_n)/[A]^n}{1 + K_1/[A] + K_1K_2/[A]^2 + \cdots + (K_1K_2\cdots K_n)/[A]^n} \tag{3.32}$$

when written in terms of the dissociation constants.

These equations are beautifully general but almost useless as they stand. If n is large, the task of fitting $\bar{v}$ or $\bar{r}$ versus [A] data to them would be formidable and very demanding of precision. To analyze the situation further, we must consider some simpler cases.

Sites Identical and Independent

If all of the sites are the same and the binding at any site is independent of the occupation of other sites, Equation (3.31) reduces to a simple form, in terms of a single constant K.

At first glance, it might seem that we need only set $K_1 = K_2 = \cdots K_n = K$ in Equation (3.31). However, there is a subtlety involved, for these constants are *not* the same for identical sites.† Note that there will be many different forms of a molecule with i sites out of n occupied. In fact, there will be

$$\frac{n!}{(n-i)!i!} \tag{3.33}$$

ways of arranging the i-occupied and $(n-i)$-unoccupied sites (see Figure 3.2).

Figure 3.2   Four of the six (4!/2!2!) arrangements of two ligands over four sites. Note that the total concentration of PA$_2$ will be the sum of the concentrations of these six forms.

The quantity [PA$_i$] in Equations (3.29) and (3.30) represents the sum of these many different concentrations. For each one we may write

$$[PA_i]_j = K^i[P][A]^i \tag{3.34}$$

where K measures the binding of A by any site. Therefore Equation (3.31) must be rewritten as

$$\bar{v} = \frac{\sum_{i=1}^{n} i \frac{n!}{(n-i)!i!} K^i[A]^i}{1 + \sum_{i=1}^{n} \frac{n!}{(n-i)!i!} K^i[A]^i} \tag{3.35}$$

† The individual constants K_i for independent binding turn out to be given by

$$K_i = K\frac{n-i+1}{i}$$

See C. Tanford (1961).

These formidable sums can be handled quite easily if it is recognized that they are related to the binomial expansion. For example, the denominator is simply $(1 + K[A])^n$. The numerator may be evaluated by noting that

$$(1 + K[A])^{n-1} = 1 + \sum_{i=1}^{n-1} \frac{(n-1)!}{(n-1-i)!i!} K^i[A]^i \tag{3.36}$$

or

$$(1 + K[A])^{n-1} = \sum_{i=0}^{n-1} \frac{(n-1)!}{(n-1-i)!i!} K^i[A]^i$$

$$= \frac{1}{nK[A]} \sum_{i=0}^{n-1} \frac{n!(i+1)}{(n-(i+1))!(i+1)!} K^{i+1}[A]^{i+1} \tag{3.37}$$

but the sum on the left is exactly the sum in the numerator of (3.35) if we replace $i + 1$ by a new index, j. (It does not matter what symbol we give the index.) Therefore

$$\bar{v} = \frac{nK[A](1 + K[A])^{n-1}}{(1 + K[A])^n}$$

$$= \frac{nK[A]}{1 + K[A]} \tag{3.38}$$

This may be rewritten in the form

$$\frac{\bar{v}}{n - \bar{v}} = \frac{\theta}{1 - \theta} = K[A] \tag{3.39}$$

Two more forms of (3.38) are useful for a graphical test for independent binding and determination of n. If we write

$$\frac{1}{\bar{v}} = \frac{1}{n} + \frac{1}{nK[A]} \tag{3.40}$$

it is clear that a graph of $1/\bar{v}$ versus $1/[A]$ should yield n and K. Alternatively, one can rearrange equation (3.40) and obtain the expression

$$\bar{v} = n - \frac{\bar{v}}{K[A]} \tag{3.41}$$

which is useful in determining n and K (see Figure 3.3). Clearly, the complicated Equation (3.31) reduces to a very simple form if the binding sites are independent and identical. Another interpretation of (3.39) is obtained by noting that

$$\frac{\theta}{1 - \theta} = \frac{[\text{Sites occupied}]}{[\text{Sites vacant}]} \tag{3.42}$$

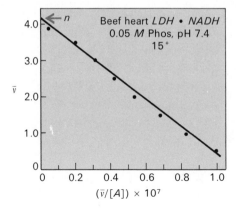

Figure 3.3

A graph according to Equation (3.41). This illustrates the binding of NADH to beef heart LDH. Evidently the binding is noncooperative, and $n = 4$. Taken from data of S. Anderson and G. Weber, *Biochemistry*. **4**, 1948 (1965). Reprinted by permission of the American Chemical Society.

where the square brackets denote "concentration of." This yields the easily recognized form

$$\frac{\text{[Sites occupied]}}{\text{[Sites vacant][A]}} = K \tag{3.43}$$

In other words, binding of this kind is indistinguishable from binding between low-molecular-weight compounds. There is simply a certain concentration of available sites in the solution and ligands to bind to them. Any reference to the macromolecular nature of the solute has vanished in (3.39). Any thermodynamic quantity calculated from (3.39) will yield the value *per site*.

Highly Cooperative Binding

Suppose that the binding of a ligand on one site of a macromolecule so strongly "activates" the other sites that they fill up immediately. In this case Equation (3.31) reduces to a very simple form, since concentrations of all species except P and PA_n will be negligible. We find

$$\bar{v} = \frac{nK[A]^n}{1 + K[A]^n} \tag{3.44}$$

or

$$\frac{\bar{v}}{n - \bar{v}} = \frac{\theta}{1 - \theta} = K[A]^n \tag{3.45}$$

The difference between independent and highly cooperative binding can easily be determined by graphing $\log\{\bar{v}/(n - \bar{v})\}$ versus $\log[A]$. Such graphs are called Hill plots. In the former case, a slope of 1 will be found and in the latter a slope of n (see Figure 3.4). However, the binding is rarely so highly cooperative as to yield a straight line with slope $n > 1$ over a wide range of $\log[A]$. A more detailed analysis shows that the graph will approach a slope of unity near the extremes† and have a maximum slope of $<n$. The behavior in the limits is understandable. After all, if there are only a very few ligand molecules, or a very few sites still available, cooperativity becomes unimportant. The binding constants determined from these limiting lines are those for the first and last molecules to be bound. See curve (c) in Figure 3.4.

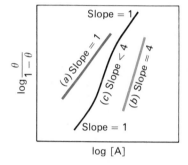

Figure 3.4
Representative Hill plots for (a) noncooperative binding, (b) highly cooperative binding with four sites, and (c) a more realistic situation for $n = 4$. See the text.

The analysis of moderately cooperative binding is difficult. Neither of the limiting forms, Equations (3.39) or (3.45), will describe the data well. Sufficiently good data can be analyzed by (3.31) if n is not too large, but the difficulties are formidable. A different approach has sometimes been employed. If we recall that equilibrium constants are measures of free-energy changes, we may attempt to include a free energy of site interaction. Formally we would write

$$\Delta G_{in}^0 = -RT \ln K_{in} \tag{3.46}$$

where K_{in} is a noncooperative *intrinsic* binding constant—that which would be observed if the site interaction were absent—and ΔG_{in}^0 is the corresponding free energy. At any finite degree of binding, $\bar{v}$, the apparent free energy will be written as the sum of ΔG_{in} and an interaction free energy, $\Delta G^0(\bar{v})$:

$$\Delta G^0 = \Delta G_{in}^0 + \Delta G^0(\bar{v}) \tag{3.47}$$

This yields, for the apparent K at $\bar{v}$,

$$K = K_{in}e^{-\Delta G^0(\bar{v})/RT} \tag{3.48}$$

† This has been used to determine the limiting constants. See Section 3.3.

It is to be understood that as $\bar{v} \longrightarrow 0$, $\Delta G^0(\bar{v}) \longrightarrow 0$ and $K \longrightarrow K_{in}$. While this looks clever, it really does not say much, for we know nothing of how $\Delta G^0(\bar{v})$ depends on $\bar{v}$. Note that $\Delta G^0(\bar{v})$ can be either negative or positive. In the former case the binding is *positively cooperative*; the presence of bound ligands facilitates further binding. If $\Delta G^0(\bar{v}) > 0$, the binding is *negatively cooperative*; previously bound ligands help keep others off. This situation is frequently encountered in ion (including proton) binding, and we shall use the formalism of (3.47) in Section 3.3.

Sets of Intrinsically Different Sites

So far, we have assumed that all of the sites on a macromolecule are intrinsically the same, though they may change in binding constant as others are filled. But many macromolecules possess groups of different kinds of sites. A common example is found in the hydrogen-ion binding by proteins, in which the various kinds of side chains (carboxylate, amino, and so forth) have very different dissociation constants. If we denote the classes of sites by I, II, and so forth, we may write

$$\bar{v} = \bar{v}_I + \bar{v}_{II} + \cdots \tag{3.49}$$

If cooperativity may be neglected, we may use Equation (3.38) for each class, giving

$$\bar{v} = \frac{n_I K_I[A]}{1 + K_I[A]} + \frac{n_{II} K_{II}[A]}{1 + K_{II}[A]} + \cdots \tag{3.50}$$

Obviously, the analysis of a binding situation of this kind will be difficult unless either the number of classes is small or the values of the K's differ greatly. We

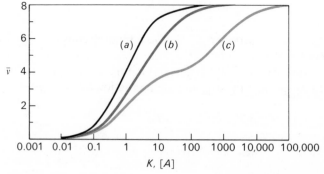

Figure 3.5   Examples of the kinds of binding curves obtained if there are two classes of sites. It is presumed that there are eight sites in all, with four in each class I and class II. Curve (a) is obtained if $K_I = K_{II}$, curve (b) if $K_I = 10\,K_{II}$, and curve (c) if $K_I = 1000 K_{II}$. In the case (b) it is difficult to tell that there are two classes.

shall consider such problems in Section 3.3. Examples of the kind of graphs obtained are shown in Figure 3.5.

3.3   TWO SPECIFIC EXAMPLES OF MULTIPLE EQUILIBRIA

In this section we shall consider as examples two important kinds of binding processes. The first, the binding of oxygen to hemoglobin, is of great physiological significance. The second, hydrogen-ion titration of proteins, is of very general interest. In neither case shall we be concerned about mechanisms of binding but rather with the analysis of the binding data.

Binding of Oxygen by Hemoglobin

The hemoglobin molecule is a tetramer, containing four oxygen-binding sites (Figure 3.6). The oxygen-binding curve does not exhibit the simple shape expected for independent binding (Figure 3.7). Independent binding *is* shown by myoglobin, an oxygen-binding protein with only one site per molecule. The

Figure 3.6   A model of the hemoglobin molecule, showing two of the heme groups that act as O_2-binding sites. The sites are roughly tetrahedrally arranged and very similar. From R. E. Dickerson, *The Proteins* (H. Neurath, ed.), Part II, 2nd Edition, Academic Press, New York, 1964.

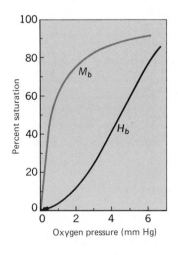

Figure 3.7
Oxygen-binding curves for human hemoglobin (four binding sites per molecule) and human myoglobin (one site per molecule). The latter conforms to Equation (3.39).

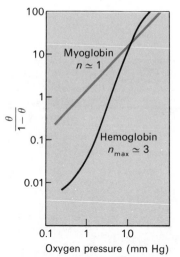

Figure 3.8
A Hill plot of the hemoglobin and myoglobin binding curves shown in Figure 3.7.

Hill graph for hemoglobin (Figure 3.8) resembles curve (c) of Figure 3.4, again indicating moderately cooperative binding.† The maximum slope depends somewhat on conditions but is always less than 4.

The most detailed studies of the equilibrium of hemoglobin with oxygen have been made by F. J. W. Roughton et al. (1955). In this work an attempt has been made to determine the individual constants in the general Adair equation [Equation (3.31)]. This can be facilitated by extremely precise measurements at very low and very high fractions of saturation, for in these cases

---

† In all of these graphs the partial pressure of oxygen is used instead of ligand concentration. This is acceptable, because the concentration of dissolved oxygen must be proportional to this pressure.

Equation (3.31) takes the limiting forms

$$\bar{v} \simeq K_1[A]/(1 + K_1[A]) \qquad \text{as} \qquad [A] \longrightarrow 0 \tag{3.51}$$

$$\simeq n - (1/K_n[A]) \qquad \text{as} \qquad [A] \longrightarrow \infty \tag{3.52}$$

Thus K_1 and K_4 may be obtained from the limiting slopes. With these constants determined, the remainder of the data may be used to fix the best values for the remaining two constants. The results, for sheep hemoglobin at pH 9.1, 19°C, are summarized in Table 3.2. The most striking aspect of these results, which must be judged as semiquantitative at best, are the large values of the enhancement factors, which provide a direct measurement of the cooperativity of the process. Hemoglobin binds the fourth oxygen molecule very much more firmly than the first molecule.

TABLE 3.2   OXYGEN BINDING BY SHEEP HEMOGLOBIN[a]

		Values of Constants, Relative to K_1		
Constant	*Predicted, Independent Binding*[b]	*Observed*	*Enhancement Factor*[c]	ΔH^d
K_1	1	1	(1)	-15.7 ± 0.8
K_2	$\frac{3}{8} = 0.375$	1.76	4.7	-11.4 ± 2.5
K_3	$\frac{1}{6} = 0.167$	1.31	7.9	-7.8 ± 3.7
K_4	$\frac{1}{16} = 0.0625$	17.7	283	-8.7 ± 3.3

[a] From Roughton, et. al. (1955).

[b] The values of the individual K_i's corresponding to independent sites. (See footnote on p. 60.)

[c] The observed constant divided by the predicted constant.

[d] Determined from the temperature dependence of the equilibrium constants [Equation (3.23)].

Recently, evidence has been developed that may help to explain these results on a molecular basis. For example, it has been shown that the distance between chains in a hemoglobin molecule changes upon oxygen binding. The role that such conformational changes may play, as well as the relation of these effects to other allosteric effects, will be discussed by Wold, in this series.

The Hydrogen-Ion Equilibrium of a Globular Protein

In this section we shall discuss the analysis of the titration of globular proteins. The example chosen is ribonuclease, for which both the amino acid sequence and the three-dimensional structure are now known. Furthermore, the very careful titration studies by C. Tanford and co-workers (1955, 1956) are some of the best available for any protein.

In discussing protein-proton equilibria, it has become traditional to use dissociation constants rather than association constants and to express the equilibrium in terms of $\bar{r}$ [Equation (3.28)] rather than $\bar{v}$. Data for the titration are shown in Figure 3.9. Several features are evident from this curve and by comparison with the amino acid composition data in Table 3.2. In the first place, not all of the 36 potentially titratable groups are titrated in the pH region from 2 to 12. It might be expected that the 4 guanidyl groups would titrate above 12, but 2 or 3 other groups are missing. Continuation of the experiments above 12 would be difficult, for irreversible changes in the protein occur in this range. Furthermore, pH measurements become very difficult at extremes of pH.

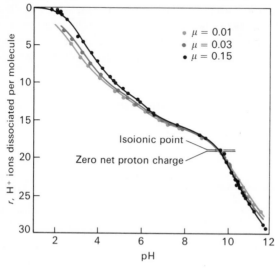

Figure 3.9   The titration of ribonuclease at 25°C. See the text. From C. Tanford and J. D. Hauenstein (1956). Reprinted with permission of the American Chemical Society.

A second complication is seen in the fact that the form of the curve evidently depends to some extent on the ionic strength of the solution. This means that we must be observing electrostatic effects in the binding, a form of cooperativity not unexpected.

Finally, although individual groups cannot be resolved, it is evident that the titration curve is roughly divided into a few ranges. This is to be expected, since the amino acid composition indicates that a limited number of types of groups are present. We should expect the carboxyl groups to have pK's ($pK = -\log K$) between 3 and 4, the histidine and the α-amino groups to titrate close to pH 7, the tyrosine and the side-chain $-NH_3^+$ groups to have pK's around 10, and the arginines to titrate above 12. This suggests that we might attempt to represent the overall titration curve by an equation analogous to (3.50):

$$\bar{r} = \frac{n_1 \mathbf{K}_1/[\mathrm{H}^+]}{1 + \mathbf{K}_1/[\mathrm{H}^+]} + \frac{n_2 \mathbf{K}_2/[\mathrm{H}^+]}{1 + \mathbf{K}_2/[\mathrm{H}^+]} + \cdots \tag{3.53}$$

Direct curve fitting to such a function would be awkward and, as we shall see, inaccurate. Other information is needed.

Some of this can be obtained by spectrophotometric titration of the tyrosines. Since the spectrum of ionized tyrosine is very different from the un-ionized form, it is possible to follow the ionization of these groups alone by measuring absorbance at 295 nm. Such data are shown in Figure 3.10. Not only do these data allow us to calculate explicitly the tyrosine contribution to Figure 3.9, but they clarify a mystery. While the data above pH 12 may not be highly accurate, Figure 3.10 shows that the six tyrosines fall into two groups; three of them titrate "normally," with a p**K** close to 10, but another three are quite abnormal, with very high p**K**. They are the three "missing groups" alluded to earlier.

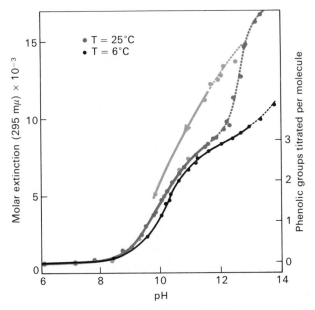

Figure 3.10 Spectrophotometric titration of the tyrosine side chains in ribonuclease. Note the two classes and evidence for irreversibility in the back titration from high pH. From C. Tanford, et al. (1956). Reprinted with permission of the Americal Chemical Society.

The assumption that each of the remaining types of groups has a p**K** value in the expected range allows separation of the curve into its component parts. However, we still do not find that the curve as predicted from amino acid composition can duplicate Figure 3.9 exactly, and we have not dealt with the ionic strength effect. To proceed further, it is necessary to take electrostatic interaction into account. Formally, this may be done in a manner analogous to (3.48). We may express each group dissociation constant in terms of an intrinsic constant, $\mathbf{K}_{i,\text{in}}$, and an electrostatic free energy of interaction, $\Delta G^0(\bar{z})$:

$$\mathbf{K}_i = \mathbf{K}_{i,\text{in}} e^{\Delta G^0(\bar{z})/RT} \tag{3.54}$$

The free energy corresponds to the work required to bring in a proton against the electrical field produced by the average charge ($\bar{z}$) on the protein molecule.

Clearly, such an analysis is approximate, for it assumes that the interaction term will depend only on the charge of the whole molecule and will be the same for all groups regardless of their locations. But it represents a start in the right direction. As a further approximation, we may assume that this energy will be proportional to $\bar{z}$. Then we may write

$$\Delta G^0(\bar{z}) = 2w\bar{z} \tag{3.55}$$

where w is an empirical parameter.† Equation (3.53) now becomes

$$\bar{r} = \frac{n_1 \mathbf{K}_{1,\text{in}} e^{2w\bar{z}}/[\mathrm{H}^+]}{1 + \mathbf{K}_{1,\text{in}} e^{2w\bar{z}}/[\mathrm{H}^+]} + \frac{n_2 \mathbf{K}_{2,\text{in}} e^{2w\bar{z}}/[\mathrm{H}^+]}{1 + \mathbf{K}_{2,\text{in}} e^{2w\bar{z}}/[\mathrm{H}^+]} + \cdots \tag{3.56}$$

If we consider the titration curve for a single type of group, we may write

$$\frac{\bar{r}_i}{n - \bar{r}_i} = \frac{x_i}{1 - x_i} = \frac{\mathbf{K}_{i,\text{in}} e^{2w\bar{z}}}{[\mathrm{H}^+]} \tag{3.57}$$

$$\mathrm{pH} - \log \frac{x_i}{1 - x_i} = p\mathbf{K}_{i,\text{in}} - 0.868w\bar{z} \tag{3.58}$$

where x_i is the fraction of groups of class i that have been titrated. This equation is exceedingly useful. A graph of the left side versus $\bar{z}$ should allow determination of both the intrinsic $p\mathbf{K}$ (from the intercept when $\bar{z} = 0$) and the value of w. Such data for the three normal tyrosine residues in ribonuclease are shown in Figure 3.11 and Table 3.3. Note the effect of ionic strength on the slope (w). As the ionic strength is increased the electrostatic effect on ionization of the protein is diminished. This is as expected and is discussed in some detail in Tanford (1961). When the electrostatic corrections are included, the predicted curves given by the solid lines in Figure 3.9 are obtained.

By proceeding in this way, it is possible to analyze and understand much of the titration curve of a protein such as ribonuclease. It should not be supposed, however, that the situation is always so straightforward. In some cases binding of ions other than protons may seriously complicate the analysis. Few proteins are conformationally stable over so wide a pH range as ribonuclease. If the dimensions of the macromolecule change with titration, w cannot be expected to remain constant, and the uptake of protons may be strongly linked to the conformational change. Such an example is seen in Figure 3.12. The abrupt change in proton binding at pH $\simeq$ 5 accompanies the transition of polyribo-adenylic acid from a single-strand to a double-strand conformation.

---

† The factor of 2 has been included to make the definition of w consistent with that used by other authors.

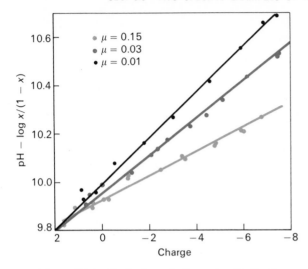

Figure 3.11 Evaluation of pK_{in} and w for the three normal tyrosines in ribonuclease. From Tanford et al. (1956). Reprinted with permission of the American Chemical Society.

TABLE 3.3   TITRATION DATA FOR RIBONUCLEASE[a]

Group	Expected pK	Number, from AA Analysis	Number Found	p$K_{i,in}$
α—COOH	~3.8	1	1[b]	—
Side chain—COOH	~4.6	10	10	4.7
Histidine	~7.0	4	4	6.5
α—NH$_2$	~7.8	1	1[b]	7.8
Tyrosine	~9.6	6	3	9.95
			3	>12
Side chain—NH$_2$	~10.2	10	10	10.2
Arginine	>12	4	4[c]	>12
Totals		36	36	

[a] From Tanford (1961).

[b] Assumed present in analyzing curve.

[c] Estimated by extrapolation of the titration curve.

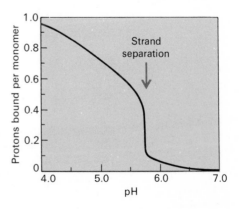

Figure 3.12
Titration of polyriboadenylic acid. Note the abrupt uptake of protons as the polymer goes over into the double-strand form at pH $\simeq$ 5. Courtesy of M. Craig.

3.4   CONFORMATION EQUILIBRIA

The equilibria between helical ordered structures and random coils appear to play an important role in biochemical processes. To cite one example, DNA appears under most circumstances as a highly ordered, double-strand helix. Yet in the processes of replication and transcription it seems very likely that this structure must be broken down, at least temporarily or in part. Again, many native proteins contain appreciable quantities of α helix. Thus it is evident that understanding of the stability of such structures is significant.

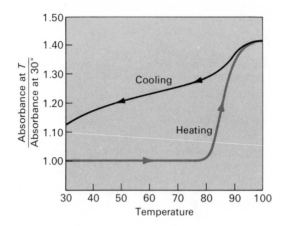

Figure 3.13
The thermal denaturation of DNA as measured spectrophotometrically. See the text.

One characteristic of many such equilibria is their *all or none* character. A typical curve for the thermal transition of DNA is shown in Figure 3.13. Below 80°C no appreciable change occurs. Then, in a very narrow temperature range, the helix "melts" to a random coil. In terms of the thermodynamics, what does this mean? If we write an equilibrium constant for the reaction as

$$K = \frac{\text{fraction of residues in random-coil regions}}{\text{fraction of residues in helical regions}} \qquad (3.59)$$

we can say

$$K = e^{-\Delta G^0/RT} = e^{-(\Delta H^0 - T\Delta S^0)/RT} \qquad (3.60)$$

A transition occurring near room temperature can be abrupt only if ΔH^0 and ΔS^0 are large, so that the exponent in Equation (3.59) changes from a very large negative quantity to a very large positive quantity for a small change in T. But what does this imply? We know that the ΔH and ΔS for the loosening of a single unit from the helix are modest quantities, probably of the order of 5 kcal and 20 entropy units, respectively. Such values would yield a gradual transition over the whole temperature range in Figure 3.13. A reasonable interpretation is as follows: The process must be highly cooperative; that is, the entire chain (or large segments thereof) must change directly from the helix to the random-coil form. If this were the case, then the observed ΔH^0 and ΔS^0 would be the sum of all of the contributions from individual pairs. DNA should then have the kind of behavior shown in Figure 3.13. Such a change is sometimes called a two-state transition.

Why should these transitions be cooperative? If one examines the structure of either an α-helical polypeptide or a helical polynucleotide, it becomes evident that it is very difficult for a single residue to move out of the helix by itself. In α helices, each residue is hydrogen-bonded to a residue four units up the chain; the bonds form an interlocking set. Similarly, a base in DNA is simultaneously hydrogen-bonded to a base on the complementary chain and interacting strongly with bases in its own chain. These, in turn, are hydrogen-bonded to the complementary chain (see Figure 3.14).

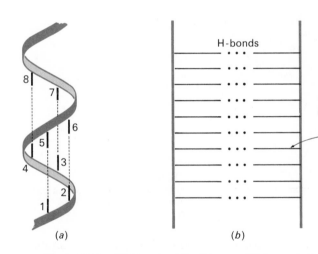

Figure 3.14 Highly schematic diagrams of (a) an α helix and (b) a double-strand poly-nucleotide. Note that in both cases three-dimensional bonding structures stabilize the structure. Contrast with the single-strand polynucleotide (c). The first two melt cooperatively and the third noncooperatively.

The same idea can be expressed in another way: It is much more difficult to start a break in the middle of a helix than it is to release a residue at the end of a helical section. Rather sophisticated statistical-mechanical treatments of this problem have been developed. We shall not go into the details here but shall simply mention some results of one such theory, that of B. H. Zimm and J. K. Bragg (1959). In analyzing this problem, these authors distinguish between the free-energy increment (ΔG_P) involved in propagating a helical region by 1 unit and that (ΔG_i) required for the initiation of a break in a helical region. The corresponding equilibrium constants are denoted by s and σ:

$$s = e^{-\Delta G_P/RT} \tag{3.61}$$

$$\sigma = e^{-\Delta G_i/RT} \tag{3.62}$$

The detailed theory shows that if $\sigma = 0$ (that is, if ΔG_i is indefinitely large) the helix-coil transition occurs at $s = 1$ with infinite sharpness. Furthermore, as σ increases, the transition becomes broader, approaching that for a noncooperative process. This limit is approached when $\sigma = 1$. For small σ, the breadth of the transition is proportional to $\sqrt{\sigma}$. Typical results from this kind of calculation are shown in Figure 3.15. The midpoint of the transition always occurs at $s = 1$ (the point where it would occur even if the transition were noncooperative). Lower values of σ simply make it sharper.

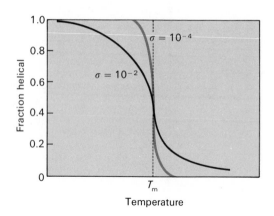

Figure 3.15
Graphs of predicted fraction helix for cooperative melting with two different values of σ.

The relationship of this to our earlier discussion is brought out by another equation:

$$\Delta H_{\text{eff}} \simeq \bar{\nu}\, \Delta H_P \tag{3.63}$$

where ΔH_{eff} is the apparent enthalpy change measured from the van't Hoff equation (3.23), ΔH_P is the enthalpy change for release of one residue, and $\bar{\nu}$ is the average number of segments that make the transition in concert.

In closing, lest it be thought that all conformational changes in macromolecules are of the cooperative type, consider the example in Figure 3.16. Polyriboadenylic acid, in neutral solution at low temperatures, forms an ordered single-strand helix. This melts gradually as shown, yielding $\Delta H^0 \simeq$

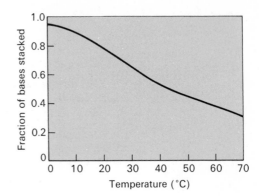

Figure 3.16
An approximate representation of the thermal denaturation of polyriboadenylic acid at neutral pH. This is a single-strand structure. See the text.

8 kcal, $\Delta S^0 \simeq 30$ e.u. These values are indistinguishable from those obtained for the dimer; thus the melting is wholly noncooperative. This should not be surprising, for the single-strand structure is stabilized only by stacking interactions of each base with its neighbors. Any such "bond" can break, without concern for the state of adjacent bonds.

TABLE 3.4 SOME CALORIMETRIC DATA ON CONFORMATIONAL TRANSITIONS[a]

Substance, reaction	pH	Solvent	(°C)	ΔH (kcal)	Method
Trypsin, denaturation	~2	0.1 M NaCl	25	8.0/mole	Heat of mixing
Myoglobin, denaturation	4.5	0.15 M KCl	30	40/mole	Heat of mixing
Ribonuclease A, denaturation	2.8	0.15 M KCl	43	70/mole	Heat capacity
Ribonuclease A, denaturation	2.8	0.15 M KCl	44	86.5/mole	Heat capacity
Fibrin, polymerization	6.08	1.0 M NaBr-acetate-phosphate	25	−19/mole	Heat of mixing
Poly-L-glutamate, denaturation	~5	0.1 M KCl	30	−1.1/residue	Heat of mixing
Poly-α-benzyl-L-glutamate, denaturation	—	25% DCE[b] 75% DCA[b]	26	0.53/residue	Heat capacity
Salmon DNA, denaturation	6.0	0.1 M NaCl	25	8.31/base pair	Heat of mixing
Sea urchin DNA, denaturation	6.0	0.1 M NaCl	25	8.03/base pair	Heat of mixing
Poly A (double-strand),	4.1	0.1 M KCl, 0.01 M cacodylate	25	2.7/nucleotide	Heat of mixing
Poly A (single-strand), denaturation	7.3	0.1 M KCl, 0.01 M tris	35	9.4/nucleotide	Heat capacity
Poly(A + U) ⟶ poly A + poly U	6.6	0.1 M KCl, 0.01 M cacodylate	25	5.9/base pair	Heat of mixing

[a] Data from H. Sober (ed.), *The Handbook of Biochemistry*, Chemical Rubber Co., Cleveland, Ohio, 1968.

[b] Abbreviations: DCE = dichloroethanol, DCA = dichloroacetic acid.

PROBLEMS

1. (a) Calculate ΔG^0 at pH 7, 37°C for the reaction

Glucose + fructose + ATP $\rightleftharpoons$ sucrose + ADP + phosphate + H_2O

(b) What is the value for the equilibrium constant?

(c) Calculate ΔG if reactants and products are all 0.001 M. In which direction does dilution drive the reaction?

2. Calculate the ratio of ADP to ATP in aqueous solution (with excess Mg^{2+}) at 37°C, as a function of total nucleotide concentration, from 0 to 1 M. Neglect the possibility of AMP formation. Assume all phosphate comes from hydrolysis of ATP and that solutions behave ideally.

3. (a) Assuming that the value of ΔG^0 for the stacking of the bases in poly A at 25°C and neutral pH is about -1.0 kcal per stacked pair, calculate the fraction of bases stacked in poly A at this temperature. You may assume the reaction to be noncooperative.

(b) Using the value of ΔH^0 given in Table 3.4 (pH 7.3), repeat the calculation at 0°C. Assume ΔH^0 and ΔS^0 are independent of temperature.

4. The data below describe the binding of a ligand A to a macromolecule. Calculate both n and the binding constant and show that all sites are identical and independent.

$[A] \times 10^3$ mole/liter	$\bar{v}$
0.5	1.6
1.0	2.5
2.0	3.2
5.0	4.0
10.0	4.1
20.0	4.8

5. The data below show the binding of oxygen to squid hemocyanin. Determine, from a Hill plot, whether the binding is cooperative and estimate the *minimum* number of subunits in the molecule.

pO_2 (mm)	% Saturation	pO_2 (mm)	% Saturation
1.13	0.30	166.8	67.3
5.55	1.33	203.2	73.4
7.72	1.92	262.2	79.4
10.72	3.51	327.0	83.4
31.71	8.37	452.8	87.5
71.87	18.96	566.9	89.2
100.5	32.9	736.7	91.3
123.3	47.8		
136.7	55.7		

6. (*a*) The molecules of a homogeneous sample of a copolymer each contain 100 aspartic acid residues ($pK \simeq 4.5$) and 50 histidine residues ($pK \simeq 7$). Neglecting electrostatic interaction, calculate the titration curve.

(*b*) Sketch an accompanying curve to show qualitatively the effects one would expect from electrostatic interaction between the sites.

7.* Consider the transition from an α helix to a random coil for a polypeptide containing n amino acid residues. Assume that the transition is cooperative.

(*a*) Assuming that one bond in each residue in the random coil may take one of *three* equiprobable orientations, obtain an equation for the number of configurations and hence the entropy change in going from 1 mole of helices (each of which has one configuration) to 1 mole of coils, each with many. (*Note:* To specify a configuration, you may regard the first two bonds as fixed.)

(*b*) If the structure (α helix) is to be stable, we must have $\Delta G = 0$ somewhere above room temperature, (say at 50°C). If the enthalpy change for breaking 1 mole of hydrogen bonds is called ΔH_{res}, obtain an expression for ΔG in terms of n and calculate what ΔH_{res} must be for $\Delta G = 0$ at 50°C, if $n = 100$. (*Note 1:* Be careful about the situation at the helix ends in relating ΔH to ΔH_{res}. *Note 2:* Experiments suggest $\Delta H_{res} \simeq +1500$ cal/mole; this gives an idea of the roughness of our approximation.)

(*c*) Using the "exact" expression for ΔG, show that the melting temperature, defined as the T for which $\Delta G = 0$, decreases with decreasing n (that is, short helices are less stable than long ones).

REFERENCES

Equilibrium Thermodynamics

Denbigh, K. G.: *The Principles of Chemical Equilibrium*, Cambridge University Press, London, 1955.

See also any good physical chemistry text.

Binding, Hydrogen-Ion Titration

Steinhart, J. and S. Beychock: in *The Proteins* (H. Neurath, ed.), 2nd ed., Academic Press, Inc., New York, 1963), Chap. 8. A broad review, with a great deal of experimental data.

Tanford, C.: *Physical Chemistry of Macromolecules*, John Wiley & Sons, Inc., New York, 1961, Chap. 8. A much more detailed treatment than given herein.

The following two articles go more deeply into the general principles and philosophy of binding studies.

Weber, G.: "The Binding of Small Molecules to Proteins," in *Molecular Biophysics* (B. Pullman and M. Weissbluth, eds.), Academic Press, Inc., New York, 1965.

Wyman, J.: "Linked Functions and Reciprocal Effects in Hemoglobin: A Second Look," *Advan. Protein Chem.*, **19**, 223 (1964).

The following papers describe aspects of the study of the hydrogen-ion equilibrium of ribonuclease.

Tanford, C. and J. D. Hauenstein: *J. Am. Chem. Soc.*, **78**, 5287 (1956).

Tanford, C., J. D. Hauenstein, and D. G. Rands: *J. Am. Chem. Soc.*, **77**, 6409 (1955).

The following paper describes in detail studies of the oxygen binding by hemoglobin.

Roughton, F. J. W., A. B. Otis, and R. L. J. Lyster: *Proc. Roy. Soc.* (*London*), **B144**, 29 (1955).

Conformational Equilibria

Birshtein, T. M. and O. B. Ptitsyn: *Conformations of Macromolecules*, Wiley-Interscience Publishers, New York, 1966. A thorough and coherent treatment of problems of molecular conformation.

Zimm, B. H. and J. K. Bragg: *J. Chem. Phys.*, **31**, 526 (1959).

FOUR | INTRODUCTION TO TRANSPORT PROCESSES: DIFFUSION

So far we have considered almost exclusively the study of systems that are at equilibrium. However, for all their rigor, such studies are quite uninformative about details of molecular structure. To find out more about the dimensions, shapes, and weights of macromolecules, the biochemist must turn to the study of systems in which transport processes are occurring, systems *not* at equilibrium.

Such processes as sedimentation, diffusion, and electrophoresis are irreversible in the thermodynamic sense; the system in which such a process is occurring is removed from a state of equilibrium. Since most of the powerful equations of classical thermodynamics apply specifically to reversible processes and systems at equilibrium, we must expect to have to do without the rigor that accompanies a thermodynamic analysis.

4.1 GENERAL FEATURES OF TRANSPORT PROCESSES

There are two very different ways in which we can attempt to analyze transport processes. We may try to see how molecules will behave, as particles subjected to external forces exerted upon them. Such a "mechanical" treatment will suffer from enormous imprecision and ambiguities, for it is obviously impossible to describe the moment-to-moment experiences of a solute molecule. Nevertheless, even a rough treatment will give a picture of what happens. Alternatively, the methods of equilibrium thermodynamics may be extended to cover some systems that are not too far from equilibrium. We may expect that in this way some of the rigor that accompanies thermodynamics may carry over, but since thermodynamics does not depend on the existence of molecules, the details of the picture will be lost. Both of these points of view will be used in succeeding chapters.

Taking at first the mechanical point of view, let us imagine in a solution a solute molecule to which an external force F_i is suddenly applied. An electric field might be switched on or a centrifugal force imposed by spinning the solution. Whatever its random motion may be, the molecule will also now be accelerated in the direction of the field. But this acceleration will last for only an *exceedingly* short time† (see Figure 4.1 and Problem 1), for as its velocity

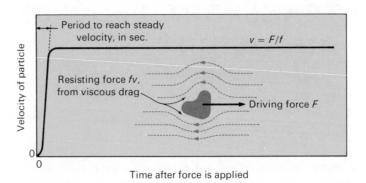

Figure 4.1   A schematic picture of the behavior of a particle suddenly subjected to a force F.

increases, the molecule will experience an increasing frictional resistance to motion through the medium. This frictional force will be given by $f_i v_i$, where v_i is the velocity and the constant f_i is called the *frictional coefficient* of the molecule.

A constant velocity will be reached when the total force on the molecule is zero:

$$f_i v_i + F_i = 0 \tag{4.1}$$

Thus, if we can measure the velocity of motion produced by a known force, we can determine the frictional coefficient. This is *one* of the reasons for the study of transport processes, for f_i depends on the size and shape of the molecule.

† The student can convince himself that this time will be of the order of nanoseconds (10^{-9} sec) by solving the equation of motion for a spherical particle falling through a viscous medium. See Problem 1.

For example, for a sphere of radius R_i

$$f_i = 6\pi\eta R_i \tag{4.2}$$

where η is the viscosity of the medium. Equation (4.2) is called *Stokes' law*; analogous equations have been worked out for particles of various shapes (see Table 4.1 and Figure 4.2).

Needless to say, this kind of analysis is very crude. We do not always know the force applied to a molecule by a known external field, and equations such

TABLE 4.1   FRICTIONAL COEFFICIENTS

Shape	Frictional Coefficient	Explanation
Sphere	$f = 6\pi\eta R$	R = sphere radius
Prolate ellipsoid	$f = 6\pi\eta R_p \dfrac{(1 - b^2/a^2)^{1/2}}{(b/a)^{2/3} \ln\{[1 + (1 - b^2/a^2)^{1/2}]/b/a\}}$	$2a$ = major axis, $2b$ = minor axis, R_p = radius of sphere of equal volume = $(ab^2)^{1/3}$
Oblate ellipsoid	$f = 6\pi\eta R_0 \dfrac{(a^2/b^2 - 1)^{1/2}}{(a/b)^{2/3} \tan^{-1}(a^2/b^2 - 1)^{1/2}}$	$2a$ = major axis, $2b$ = minor axis, R_0 = radius of sphere of equal volume = $(a^2b)^{1/3}$
Long rod	$f = 6\pi\eta R_r \dfrac{(a/b)^{2/3}}{(3/2)^{1/3}\{2 \ln[2(a/b)] - 0.11\}}$	a = half-length, b = radius, R_r = radius of sphere of equal volume = $(3b^2a/2)^{1/3}$

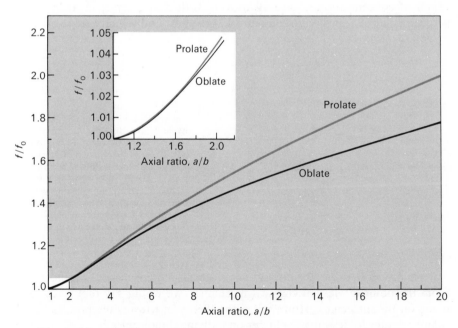

Figure 4.2   The dependence of frictional coefficient on particle shape. The ratio f/f_0 is the frictional coefficient of an ellipsoid of the given axial ratio divided by the frictional coefficient of a sphere of the same volume as the ellipsoid.

as Stoke's law have been derived only for the motion of particles through continuous media. The actual bumping and jostling of molecules is neglected entirely. However, the fact that equations such as (4.2) depend on molecular dimensions is one reason for the importance of transport processes in the study of macromolecules (see Section 5.1).

A much more elegant framework for the description of transport processes has been developed in recent years. This is the *thermodynamics of irreversible processes*, which allows the discussion of systems not too far from equilibrium in much the manner of classical thermodynamics. We shall not attempt to develop this theory here but shall use some of its methods to give a more exact and coherent picture of transport processes. There is an altogether different reason the biologist and biochemist should be acquainted with this theory. The life of cells and organisms is an irreversible process, at all levels.

The phenomena that we shall discuss (diffusion, sedimentation, electrophoresis, and so forth) have one feature in common: A system *not* in equilibrium moves towards equilibrium. The final state is dictated, of course, by the temperature, pressure, composition, and external forces imposed on the system. During the drive toward equilibrium, *flow* must occur. In the problems of interest to us, this is a flow of matter—as, for example, in the movement of molecules in a centrifugal field. We define the flow (J_i) of a component i as the number of mass units (grams, moles, and so forth) of i crossing 1 cm^2 of surface in 1 sec. The flow is obviously related to the velocity of transport. If solute molecules in a region of concentration C_i are all moving with a velocity v_i in a direction perpendicular to a surface s (see Figure 4.3), the flow and velocity are related by

$$J_i = v_i C_i \tag{4.3}$$

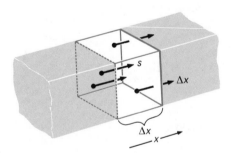

Figure 4.3

The relation between molecular velocity and flow. In time Δt a molecule with velocity v moves a distance $\Delta x = v\Delta t$. Thus, all the solute molecules in a slab going back this distance from the surface s will pass through s in Δt seconds. The amount passing will be the concentration times the slab volume $\Delta w = C \cdot s\Delta x$, or $\Delta w = C \cdot sv\Delta t$. Since the flow is defined as $J = \Delta w/s\Delta t$, we obtain $J = Cv$.

What, precisely, drives the flow when a system is moving toward equilibrium? From the point of view of classical mechanics, we would say that forces are acting on the molecules. However, in the theory of irreversible processes, we do not attempt to describe the forces on individual molecules but take a more general view. In classical mechanics, a force is always the negative of the gradient (rate of change with distance) of a potential energy; similarly we shall identify the generalized "forces" that drive the processes of diffusion, sedimentation, and electrophoresis with the gradients of certain "potentials." When these potentials become uniform throughout the system, no force exists, and flow no longer occurs.

An example is familiar to the reader: If potential differences exist in an electrical circuit, current flows until these differences are eliminated. The similarity of this and the processes we are discussing is pointed up in Table 4.2.

TABLE 4.2   TRANSPORT PROCESSES

Process	Potential	Flow of	State of Equilibrium
Electrical conduction	Electrostatic	Electrons	Uniform electrostatic potential
Heat conduction	Temperature	Heat	Uniform temperature
Diffusion	Chemical potential	Molecules	Uniform chemical potential
Sedimentation	Total potential = chemical potential + centrifugal potential energy	Molecules	Uniform total potential

A primary result of the theory of irreversible processes is the statement that the flow, at any point in the system, at any instant, is proportional to the gradient in the appropriate potential, there and then. For one-dimensional flow in an ideal system,

$$J_i = -L_i \frac{\partial U_i}{\partial x} \tag{4.4}$$

This equation can be thought of as a generalization of Ohm's law with U_i a generalized potential and L_i a generalized "conductivity."[†] It says that the farther the system is from equilibrium, the faster it moves toward it. Equation (4.4) is written for potentials that change in one direction only; the extension to three dimensions is easy, but then both the gradient and the flow *must* be written as vector quantities. We can now connect the mechanical and thermodynamic analyses of transport phenomena. If we regard the force per molecule F_i in (4.1) as $-(1/\mathcal{N})(\partial U_i/\partial x)$, where $\mathcal{N}$ is Avogadro's number, and recall Equation (4.3), then (4.4) may be rewritten as

$$C_i v_i = \mathcal{N} L_i F_i \tag{4.5}$$

or

$$v_i = \frac{\mathcal{N} L_i}{C_i} F_i \tag{4.6}$$

which says that $L_i = C_i/\mathcal{N} f_i$, $\mathcal{N} f_i$ being the frictional coefficient per mole.[‡]

[†] Equation (4.4) is for a relatively simple case, where one kind of potential gradient and one kind of flow exist. If one has gradients in several potentials and flow of several quantities (several chemical components, heat, and so forth), the general result is $J_i = -\sum_k L_{ik}(\partial U_k/\partial x)$. This says that the flows may be *coupled*; the existence of a gradient in potential k may cause a flow of substance i.

[‡] The relationship $L_i = C_i/\mathcal{N} f_i$ may be thought of in the following way: It says that the *mass conductivity* (L_i) is proportional to the concentration of the substance being transported and inversely proportional to the resistance the medium offers to that transport.

It is well to remember that the two theories start from different points of view and that the f_i defined from the thermodynamic theory may not be *exactly* the same quantity that we would obtain from a naive calculation from molecular dimensions.

Before turning to detailed description of particular transport processes, there is one more general point to consider. The experimenter rarely measures flow directly; more often, the changes in concentration with time in various parts of the system are followed. Thus we must relate flow to these concentration changes. This can be accomplished by utilizing the fact that matter is not being created or destroyed in the system, in other words, the conservation of mass. A cell of uniform cross section (A) in which a flow of solute is occurring is shown in Figure 4.4. Consider a thin slab, perpendicular to the direction of flow, with

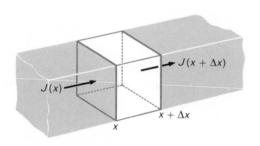

Figure 4.4

The continuity equation. The change of concentration in the slab between x and $x + \Delta x$ depends upon the difference between the flow in and the flow out. See text.

surfaces at x and $x + \Delta x$. How will the concentration change in this volume element in time Δt? In the general case, we shall assume that J_i may be different at different points. Then, the net change in the mass of solute in the slab (Δw_i) must be given by the flow *in* [$J_i(x)$] minus the flow *out* [$J_i(x + \Delta x)$], each multiplied by the area and Δt:

$$\Delta w_i = J_i(x)A\,\Delta t - J_i(x + \Delta x)A\,\Delta t \tag{4.7}$$

If we divide this by the volume of the slab, we obtain the change in concentration:

$$\Delta C_i = \frac{\Delta w_i}{A\,\Delta x} = \frac{J_i(x) - J_i(x + \Delta x)}{\Delta x}\Delta t \tag{4.8}$$

or, rearranging,

$$\frac{\Delta C_i}{\Delta t} = \frac{J_i(x) - J_i(x + \Delta x)}{\Delta x} \tag{4.9}$$

When the increments become infinitesimal, we get the differential equation of continuity

$$\frac{\partial C_i}{\partial t} = -\frac{\partial J_i}{\partial x} \tag{4.10}$$

In words, this simply says that the concentration will increase in a region if material is coming in faster than it is going out. Obviously, the equation will have to be modified if chemical reactions are making or using up substance i or if A depends on x.

4.2 DIFFUSION

Let us now apply the generalities that have been developed to a specific transport process—diffusion. The arrangement shown in Figure 4.5 is one way in which diffusion can be studied. By one or another mechanical trick, solvent is layered carefully on a solution. This gives a sharp *boundary*, across which the solute concentration changes abruptly from zero to a value C_0. Common sense indicates that this situation cannot persist, for the random Brownian motion of the molecules will blur the boundary and eventually lead to complete mixing.† Thermodynamics says the same thing in a different way; the initial state is certainly not one of equilibrium, for the free energy will be lower when uniform mixing has occurred. Neglecting small contributions from the heat of mixing, the free-energy change results entirely from the higher entropy of the mixed state.

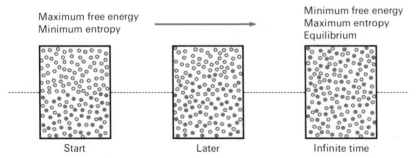

Maximum free energy
Minimum entropy

Minimum free energy
Maximum entropy
Equilibrium

Start   Later   Infinite time

Figure 4.5   The process of diffusion. Open circles are solvent molecules; filled circles are solute.

In fact, our previous discussion of the thermodynamics of solutions gives a very precise prescription for the equilibrium state: The chemical potential must be constant throughout the system. Thus the change of chemical potential with distance in the boundary region is the "driving force" for the flow of solute that will lead to equilibrium:

$$J_2 = -L_2 \frac{\partial \mu_2}{\partial x} \tag{4.11}$$

† It is possible to derive the equations for diffusion from a consideration of Brownian motion as a random-walk process. The analysis is much more difficult and leads to somewhat ambiguous results for any but the most simple systems. For example, it is almost impossible to properly include effects of solution nonideality.

The subscript 2 denotes the solute component. The chemical potential of the solute depends on temperature, pressure, and the concentration, but since T and P are assumed to be uniform,

$$\frac{\partial \mu_2}{\partial x} = \left(\frac{\partial \mu_2}{\partial C_2}\right)_{T,P} \frac{\partial C_2}{\partial x} \tag{4.12}$$

Thus under these conditions the concentration gradient determines the flow. From Equation (2.33), $(\partial \mu_2/\partial C_2)_{T,P} = (RT/C_2)[1 + C_2(\partial \ln y_2/\partial C_2)]$, where y_2 is the activity coefficient. Then

$$J_2 = -\frac{L_2 RT}{C_2}\left\{1 + C_2\frac{\partial \ln y_2}{\partial C_2}\right\}\frac{\partial C_2}{\partial x} \tag{4.13}$$

The factor L_2/C_2 in Equation (4.13) can be replaced by $1/\mathcal{N}f_2$, bringing in the frictional coefficient:

$$J_2 = -\frac{RT}{\mathcal{N}f_2}\left\{1 + C_2\frac{\partial \ln y_2}{\partial C_2}\right\}\frac{\partial C_2}{\partial x} \tag{4.14}$$

or

$$J_2 = -D_2\frac{\partial C_2}{\partial x} \tag{4.15}$$

where

$$D_2 = \frac{RT}{\mathcal{N}f_2}\left\{1 + C_2\frac{\partial \ln y_2}{\partial C_2}\right\} \tag{4.16}$$

The quantity D_2 is termed the *diffusion coefficient*. From an experimental point of view it is defined by Equation (4.15), called *Fick's first law*. This equation is in accord with our intuitive expectation that the flow will cease only when the concentration is uniform. Equation (4.16) asserts that D depends on RT (which may be taken as a measure of the kinetic energy of the molecules); the correction term $[1 + C_2(\partial \ln y_2/\partial C_2)]$, which expresses the fact that the chemical potential depends on solute-solute interaction, and the size and shape of the molecule, which will influence the frictional coefficient. It is the last quantity that makes the macromolecular chemist interested in diffusion. For ideal solutions we may neglect the activity coefficient factor and obtain $D_2 = RT/\mathcal{N}f_2$. If, in addition, f_2 does not vary with concentration, D_2 becomes a constant.

Equation (4.15) does not suggest an easy method for the determination of D.† We would rather know how the concentration changes with time at various

† Actually, D can be determined by direct use of Equation (4.15). See Problem 3.

points in the boundary. This can be obtained by combining (4.15) with the continuity equation (4.10):

$$\left(\frac{\partial C_2}{\partial t}\right) = -\left(\frac{\partial J_2}{\partial x}\right) \quad \text{and} \quad J_2 = -D_2\left(\frac{\partial C_2}{\partial x}\right)$$

$$\left(\frac{\partial C_2}{\partial t}\right) = D\left(\frac{\partial^2 C_2}{\partial x^2}\right) \tag{4.17}$$

Here the assumption has been made that D does not change with concentration, usually a satisfactory approximation.

Equation (4.17), which is called *Fick's second law*, is very much like the partial differential equation for heat conduction. It may be solved for specific cases if we specify initial and boundary conditions. Figure 4.6 shows the most common

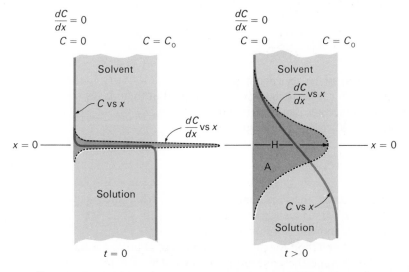

Figure 4.6 The spreading of an initially sharp boundary during free diffusion. The heavy line traces the concentration as a function of height in the cell (Equation 4.18). The broken line shows the concentration gradient, according to Equation 4.19.

experimental arrangement for diffusion studies. The solution column is quite long, so that it may be assumed that the concentrations at the top and bottom ($x = +\infty$ and $-\infty$) remain zero and C_0, respectively, as long as the process is followed. A sharp boundary is formed at $x = 0$ at time $t = 0$ (see Chapter 6 for the forming of such boundaries). Under these conditions of *free diffusion*, the solution of Equation (4.17) is

$$C_2 = \frac{C_0}{2}\left\{1 - \frac{2}{\sqrt{\pi}}\int_0^{x/2(Dt)^{1/2}} e^{-y^2}\, dy\right\} \tag{4.18}$$

where y is simply a dummy variable of integration. This equation gives the kind of S-shaped curve shown in Figure 4.6 (see Problem 5).

Equation (4.18) is awkward to use or visualize, but its derivative (the concentration gradient) is simply a Gaussian "error" curve:

$$\left(\frac{\partial C_2}{\partial x}\right) = \frac{C_0}{2(\pi Dt)^{1/2}} e^{-x^2/4Dt} \tag{4.19}$$

It turns out that the gradient is easily measured by use of a schlieren optical system (see Section 4.3). The height of the bell-shaped curve is given by

$$\left(\frac{\partial C_2}{\partial x}\right)_{x=0} = \frac{C_0}{2(\pi Dt)^{1/2}} = H \tag{4.20}$$

and the area under the curve is simply C_0:

$$C_0 = \int_{-\infty}^{\infty} \left(\frac{\partial C_2}{\partial x}\right) dx = A \tag{4.21}$$

The latter result can be seen from the fact that $(\partial C_2/\partial x)\,dx$ is the increment in concentration for a particular dx; summing all of these gives the total concentration difference across the boundary.

Combining (4.20) and (4.21), we have

$$\frac{A}{H} = 2(\pi Dt)^{1/2} \tag{4.22}$$

As diffusion proceeds, the area remains constant, while the height decreases. A graph of $(A/H)^2$ versus t will give a straight line with slope $4\pi D$. An example of such a graph is shown in Figure 4.7.

A more accurate method for the measurement of D utilizes a Rayleigh interferometer to determine the concentrations at different points in the cell (see Section 4.3). This requires, of course, the use of Equation (4.18). Essentially,

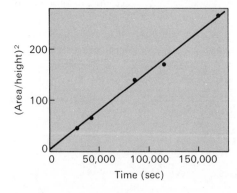

Figure 4.7

Determination of the diffusion coefficient according to Equation 4.22. The fact that the line does not pass exactly through the origin is not surprising; the boundary was not perfect at the beginning of the experiment, so the data behave as if diffusion had already started. The solute is ovalbumin; data of Lamm and Polson, *Biochem. J.*, **30**, 528 (1936).

the calculation proceeds as follows: A particular concentration, at a specific time, corresponds to a specific value of $x/2(Dt)^{1/2}$. The value of x at which this concentration exists can be measured from the position of a given fringe in the interference pattern. Thus D is determined at a whole series of points at the time of each photograph. By this procedure, average values of D with a precision of about ± 0.1 percent can be determined. An even more accurate interferometric technique called the Gouy method uses the symmetrical boundary itself to produce an interference pattern. So far, this method has been used mainly for low-molecular-weight materials, since interpretation of the patterns is easy only for rigorously Gaussian boundaries. Some results obtained by these methods are summarized in Table 4.3.

TABLE 4.3   DIFFUSION COEFFICIENTS

Substance	Molecular Weight	$D_{20,w} \times 10^6$	Method
Glycine*	75	9.335	G
Sucrose*	342	4.586	G
Ribonuclease	13,683	1.068	R
Serum albumin* (bovine)	66,500	0.603	R
Tropomyosin	93,000	0.224	S
Fibrinogen* (human)	330,000	0.197	R
Myosin*	440,000	0.105	S
TMV	About 40,000,000	0.053	S
Rabbit papilloma virus	About 47,000,000	0.059	S

[a] The diffusion coefficients have been corrected to water at 20°C (see Chapter 5 for the procedure). Those indicated by an asterisk have been extrapolated to zero concentration. The methods are G, Gouy; R, Rayleigh; and S, Schlieren. Note that D generally decreases with molecular weight but that elongated molecules such as tropomyosin, fibrinogen, myosin, and TMV have unusually low values. The dimensions of D are $cm^2 \cdot sec^{-1}$.

A final word of caution is in order concerning diffusion measurements and transport processes in general. The equations most often used are derived from systems involving only two components, solute and solvent. A single transport coefficient (such as L or D) defines the flow for such systems. If there are more than two components, *interaction of the flows* may occur. In diffusion, for example, this means that a concentration gradient in one component causes diffusion of the other. (After all, the chemical potential of a component depends in general on the concentrations of *all* solutes.) The theory of irreversible processes shows that, in general, a three-component system will require *four* diffusion coefficients. The analogs of Fick's first law become

$$J_2 = -D_{22}\frac{\partial C_2}{\partial x} - D_{23}\frac{\partial C_3}{\partial x}$$

$$J_3 = -D_{32}\frac{\partial C_2}{\partial x} - D_{33}\frac{\partial C_3}{\partial x} \tag{4.23}$$

where 2 and 3 denote solute components. In general, one expects, and finds, that D_{22} and D_{33} are close to the coefficients that would be observed if each component diffused alone and that the cross-term coefficients D_{23} and D_{32} are small. However, the effects may be appreciable, especially if the concentration of one of the components is large. Furthermore, it should be pointed out that biochemists rarely work with two-component mixtures; even a highly purified protein is generally studied in the presence of buffers and neutral salts.

4.3   OPTICAL SYSTEMS FOR THE STUDY OF TRANSPORT PROCESSES

The quantitative study of transport processes such as diffusion, sedimentation, and electrophoresis requires that we be able to follow changes in the shape and/or position of a boundary without disturbing the system. Solutions to this problem have almost invariably been based on optical methods. In this section we shall describe those methods of widest use. Understanding of their operation is essential to intelligent use of these techniques in research and even to critical analysis of results in the literature. In keeping with the content of this chapter, we shall describe the methods as they might be used to follow free diffusion, but they can be adapted with little change for use in ultracentrifugation and electrophoresis.

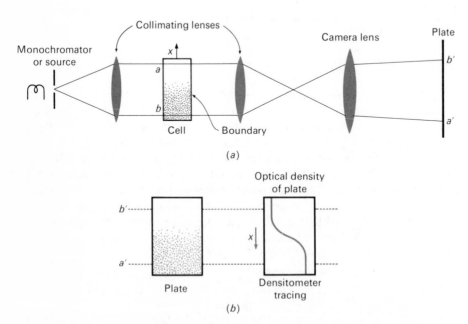

Figure 4.8   (a) An absorption optical system. An inverted image of the cell is focused on the photographic plate by the camera lens. (b) A representation of the photographic plate that might be obtained from a diffusion experiment as in (a), and the densitometer tracing which might be obtained from it.

Absorption Optical Systems

The simplest method for determining the concentration of a solute at different points in a cell is to take advantage of the absorption of light by the substance. In most cases this means utilizing ultraviolet radiation; this may or may not be monochromatic. If it is monochromatic, one may be able to measure the concentrations of two or more solutes independently by changing the wavelength to correspond to specific absorption bands. In either case, the most common arrangement is shown in Figure 4.8(a). Light from the source is passed through the cell, and a camera lens focuses an image of the cell onto a photographic plate (Figure 4.8(b)). The darkening of the plate at various levels in the cell image will depend on the amount of light that can pass through each level. By scanning the photographic plate with a densitometer, this blackening can be measured and, on proper calibration, will yield a concentration versus distance curve. The calibration is somewhat awkward, since the blackening of the plate will depend on exposure time, development, and characteristics of the plate, as well as the optical density of the solution. A technique that avoids some of these problems has recently been used in the ultracentrifuge; the photographic plate is replaced by a traveling slit and photomultiplier assembly, which scans the image directly.

Schlieren Optical Systems

Schlieren optical systems are the most widely used at the present time, especially for sedimentation velocity studies. All schlieren optical systems depend on the following fact: When parallel light passes through a region where there is a refractive index gradient, it is deflected as by a prism (see Figure 4.9). It can be shown that the deflection angle is given approximately by

$$\alpha = a(dn/dx) \tag{4.24}$$

where a is the cell thickness and dn/dx is the refractive index gradient. Thus parallel light passing through a cell in which there is a boundary will be deflected only in the boundary region, and most at the steepest part of the gradient.

Figure 4.9

The passage of parallel light rays through a boundary. The wave fronts of the incident light can be visualized as planes perpendicular to the direction of propagation of the rays. The light is retarded more in the high concentration (high refractive index) part of the cell. This results in distortion of the emergent wave front (greatly exaggerated here), or, in other terms, deflection of rays which pass through the boundary. Note that the rays are deflected most where the refractive index gradient is greatest.

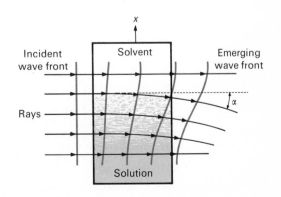

A method that measures refractive index gradient also measures, in effect, concentration gradient, since for a dilute solution the refractive index difference between solution and solvent is directly proportional to the concentration.

To understand the operation of modern schlieren systems, it will be useful to consider first a more primitive version, which is no longer used [Figure 4.10(a)]. It is arranged very much like the absorption system in Figure 4.8(a). However, in tracing rays coming through the cell at various levels, we note that deflected rays from regions of equal gradient (a and a', for example) are focused below the optic axis at plane P. Of course, they will normally reach the photographic plate at the correct level on the cell image; the camera lens take care of that by collecting all bundles of rays from a given point in the cell and focusing them at a corresponding point on the cell image.

But now suppose we place a knife edge at plane P and begin to raise it toward the optic axis. The first rays intercepted will be those the most deflected, those coming through the region of maximum gradient. Now they will not reach the plate, and the image will be dark at the center of the boundary. As

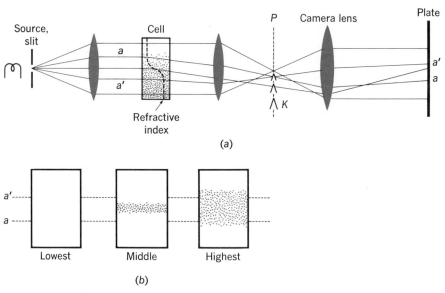

Figure 4.10   (a) A primitive schlieren system. Rays which have been deflected by gradients in the cell pass the plane P below the optic axis. If a knife edge (K) is placed in this plane, it will intercept such rays. (b) The cell image as seen with each of the three knife edge positions in (a). In the lowest position, no rays are intercepted. In the second, those coming from the steepest part of the boundary region have been cut off; a shadow appears in the corresponding part of the cell image. In the third, the edge is so close to the optic axis that the shadow extends over most of the boundary region.

the knife edge is raised more and more, the dark band will become broader [see Figure 4.10(*b*)]. Finally, as the knife edge cuts the optic axis, the whole cell image will go dark.

The above technique will locate a boundary but is a pretty awkward way to study its shape. Therefore, let us turn to the schlieren system as it is now used. Shown in perspective [Figure 4.11(*a*)] this looks more complex than the earlier system. Actually, only two elements have been added: a plate with a diagonal slot at the focal plane *P* and a cylindrical lens. The cylinder lens is so placed that in the direction *perpendicular* to the *x* axis of the cell, plane *P* is focused at the plate. Along the cell axis direction, the cylinder lens does not contribute, so the cell itself is in focus at the plate in this direction. The system is very astigmatic. To see what happens, recall the primitive system. It might be said that a whole series of slit images were formed at plane *P*, the one on the optic axis being formed from rays passing through zero gradient regions,

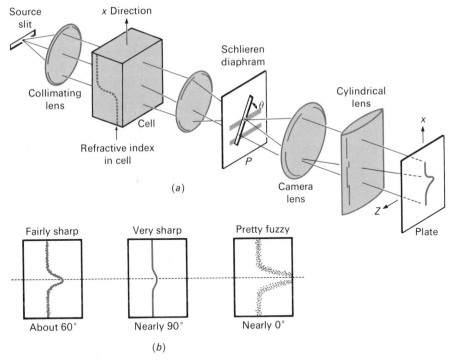

(*a*)

(*b*)

Figure 4.11 (*a*) The cylindrical lens schlieren system. In the form pictured, there is a diagonal slot at the focal plane *P*. One can imagine light passing through levels of different refractive index gradient forming slit images at *P* which are displaced to various distances below the optic axis. Since the cylindrical lens focuses the plane *P* on the plate (in the *Z* direction only), the fact that off-axis light can come through the diagonal slot only to one side yields a *Z*-deflection on the plate proportional to the refractive index gradient. (*b*) Schlieren images as would be produced by the system shown in (*a*) with various settings of the angle *θ*.

the one farthest off axis being formed from rays passing through the maximum gradient region. The diagonal slit has this function: The farthest deviated rays can pass through it only to one side on the plane P (the slit is diagonally placed). Now consider the fate of any ray that passes through the system. No matter how much it is deviated, it will get back to the proper height on the cell axis (the camera lens takes care of this, and the cylinder lens does not interfere). But the cylinder lens does focus the *sideways* deflections at plane P on the plate. Therefore a deviated ray will end up with a sideways deflection on the plate, compared to an undeviated ray. Thus a profile, with sideways deflection proportional to the downward deflection at the cell, will be traced out on the plate. But the downward deflection is just proportional to the refractive index gradient. The sideways deflection will depend on this, on the angle of the diagonal slit (θ), and on the magnification factor of the cylinder lens (G). In fact, we find

$$Z = Gab(\cot \theta)\, dn/dr \qquad (4.25)$$

The angle θ can be varied to change the deflection Z corresponding to a given gradient. Typical results are shown in Figure 4.11(*b*).†

The schlieren system is elegantly suited to examine boundary shape, and it has been widely used. It is not, however, the most accurate method; usually one cannot obtain precise results with concentrations less than a few milligrams per milliliter.

Interferometric Optical System

The application of interferometry to the study of transport processes has increased the accuracy of some measurements by nearly one order of magnitude. Most commonly employed is the Rayleigh interferometer shown in Figure 4.12(*a*). The arrangement is much like that in the cylindrical lens schlieren systems. However, the source slit is rotated by 90 deg, to be parallel to the cell axis. A double cell is used, with solvent in one side and the solution-solvent boundary in the other. If a pair of slits are inserted, one behind each channel, a double slit interference pattern might be expected at the focal plane P. However, the actual pattern at this plane will be much more complex, because of the bending of light and because the optical path difference is different at various heights in the double cell. To sort this out, the same kind of astigmatic system employed in the schlieren system is used. Once more, the cell is in focus at the plate *along the cell axis*; but at right angles to this, because of the cylindrical lens, the focal plane P is in focus. What happens can best be imagined by visualizing the cell as divided into a series of layers, perpendicular to its axis. Each layer has "its own" interference pattern, with fringes shifted accord-

---

† In actual practice, a half-wave phase boundary plus a fine metal line are frequently used instead of a diagonal slit. This gives a dark line on a light field, instead of a light line on a dark field, as would the arrangement described above.

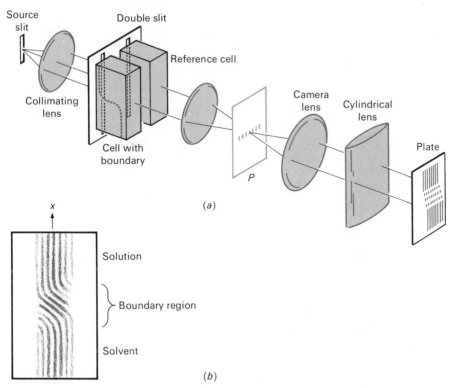

Figure 4.12   (*a*) A schematic diagram of the Rayleigh interference system. Each layer in the cell can be thought of as providing its own interference pattern. These are all superimposed at *P*, but are sorted out by the astigmatic optical system. The image shown on the plate is intended to illustrate this idea. Actually, the pattern will be similar to that shown (*b*). (*b*) The kind of fringe pattern produced by a boundary in the Rayleigh system. Only the central diffraction band of the image shows up; superimposed on this are the Rayleigh fringes.

ing to the refractive index difference between the two channels at that layer. Each layer is, of course, focused at the correct height. The overall pattern obtained is shown in Figure 4.12(*b*). The progressive shifting of the fringes as they pass through the boundary region means that each fringe traces the refractive index versus distance curve. The refractive index difference (Δn) between two points on the cell axis can be judged by the number of fringes crossed (Δj) in going between these two points:

$$\Delta j = \frac{a\,\Delta n}{\lambda} \tag{4.26}$$

Here *a* is the cell thickness and λ the wavelength of the light used. This makes measurement of refractive index (and hence concentration) difference very easy and accurate.

PROBLEMS

1. Consider a small spherical particle of mass m, initially at rest, which is acted upon by a constant force F turned on at $t = 0$. Solve the equation of motion for this particle, which must be $F - fv = m\,dv/dt$, where v is the velocity. Show that it approaches a constant velocity $v_{max} = F/f$. If the particle is spherical, of radius 100 Å, and of density 1.5 g/cm^3, calculate the time required to attain 99 percent of the final velocity.

*2. Two large compartments, A and B, are separated by a porous disk or membrane of thickness Δx, through which a solute can diffuse. The solute is initially present at different concentrations, C_A and C_B, in the two compartments. These are stirred, and diffusion is allowed to proceed through the disk. If A and B are large, a steady state will be obtained *in the disk* in which concentration will not change with time. Show that the gradient in the disk will be linear:

$$C(x) = C_A + (C_B - C_A)x/\Delta x$$

3. Using the result of Problem 2 and Fick's first law, derive an expression for the change of $C_B - C_A$ with time, assuming that A and B each have volume v and that the porous disk has an effective area s. This is the principle of the porous-plate method for measurement of D.

4. Obtain an equation for the half-width of a free diffusion boundary—the value of x for which $(\partial C/\partial X) = H/2$. Calculate the half-width of boundaries of sucrose and TMV after 1-hr diffusion at 20°C.

*5. Obtain Equation (4.19) and then (4.18) by solution of Fick's second law [Equation (4.17)] subject to the boundary conditions $C = 0(x = +\infty)$, $C = C_0(x = -\infty)$. [*Hint:* This is difficult unless you make the guess that the solution will be of the form $C = C(u)$, where $u = x/t^{1/2}$. Choice of this new variable reduces the partial differential to a simple differential equation in u. Another substitution, $y = dc/du$, allows easy integration.]

6. Frictional coefficient data are frequently informative about molecular shape. From the data given in Table 4.3, calculate the frictional coefficients (per molecule) for serum albumin and tropomyosin. Now, presuming that the molecular weights are correct and that the specific volume of each protein is about 0.74 cm^3/g, calculate the volume of each molecule and the radius it would have if it were an unhydrated sphere. Using this radius, together with Stoke's law, you can calculate the minimum value of the frictional coefficients that these proteins could have. How do these compare with the observed values? If you assume each molecule is a prolate ellipsoid, what axial ratios are found? Now assume that each molecule is really spherical but highly hydrated. What radius, volume, and hydration (grams of water per gram of protein) are required to account for the observed frictional coefficients? Are these values reasonable? What can one conclude?

*7. Using the primitive schlieren system in which one simply observes a black band corresponding to a boundary, devise a way for measuring D in free diffusion. Work out the equations and show what kind of graph you would draw to calculate D.

8. How many Rayleigh fringes would one count across a boundary between a 1-mg/ml solution of a protein and solvent if the cell thickness (a) is 3 cm and the specific refractive index increment (dn/dc) is 0.186 (g/ml)$^{-1}$? Assume that monochromatic light of $\lambda = 5460$ Å (Hg green line) is used.

REFERENCES

General

Katchalsky, A., and P. F. Curran: *Nonequilibrium Thermodynamics in Biophysics*, Harvard University Press, Cambridge, Mass., 1965.

Diffusion

*Einstein, A.: *Investigations on the Theory of the Brownian Movement*, Dover Publications, Inc., New York, 1956. A classic, picturing diffusion in a rather different way than we have used.

Gosting, L. J.: *Advan. Protein Chem.*, **11**, 429 (1956). A very thorough review of experimental methods and interpretations.

FIVE | SEDIMENTATION

The development of the ultracentrifuge, in the 1920s, marked in one sense the beginning of molecular biology. For the first time it was possible to un-ambiguously measure the molecular weights and heterogeneity of macro-molecular substances. Now this instrument is used routinely for the separation and purification of macromolecular substances, the analysis of mixtures, and the determination of molecular weights and dimensions. In fact, the applications of centrifugal methods in biochemistry are so diverse that we can only touch on them here; for more detail about theory and practice, the reader should turn to the references given at the end of the chapter.

5.1 SEDIMENTATION VELOCITY

We shall introduce the concept of sedimentation by a simple mechanical analysis, leaving the more mathematically elegant theory for later. Imagine a solute molecule in a solution that is held in a rapidly spinning rotor. Forgetting, for the moment, the random pushes and pulls that the molecule receives from its neighbors, we may say that there are three forces acting on it (see Figure 5.1). If the rotor turns with an angular velocity ω (radians per second), the molecule will experience a *centrifugal* force proportional to the product of its mass (m) and the distance (r) from the center of rotation: $F_c = \omega^2 rm$. At the same time, the molecule displaces some solution; this will give rise to a force equal to that which would be exerted on the mass of solution displaced: $F_b = -\omega^2 rm_0$. Finally, if as a result of these forces the molecule acquires a velocity (v) through

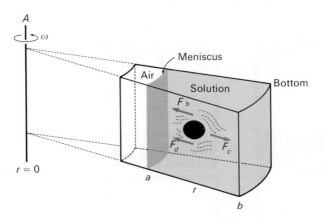

Figure 5.1   Diagram of a sedimentation experiment (not to scale). The sector-shaped cell is in a rotor spinning about the axis A at an angular velocity ω. The molecule is acted on by centrifugal, buoyant, and frictional drag forces. The cell has been given a sector shape because sedimentation proceeds along radial lines; any other shape would lead to concentration changes near the edges, with accompanying convection. The angular velocity is given by $(2\pi/60)$ times revolution per minute.

the solution, the kind of viscous drag discussed in Chapter 4 will be experienced: There will be a *frictional force* $F_d = -fv$, where f is the frictional coefficient. As indicated in Chapter 4, this kind of situation will result in the molecule acquiring a velocity just great enough to make the total force zero:

$$F_c + F_b + F_d = 0$$

$$\omega^2 rm - \omega^2 rm_0 - fv = 0$$

(5.1)

For the mass of solution displaced, m_0, we may substitute the product of the particle mass times its partial specific volume, times the solution density†: $m_0 = m\bar{v}\rho$. Therefore,

$$\omega^2 rm(1 - \bar{v}\rho) - fv = 0$$

(5.2)

Next, we multiply Equation (5.2) by Avogadro's number, to put things on a mole basis, and rearrange, placing molecular parameters on one side of the equation and experimentally measured ones on the other:

$$\frac{M(1 - \bar{v}\rho)}{\mathcal{N}f} = \frac{v}{\omega^2 r} = s$$

(5.3)

† At this point the crudeness of our model has caught us up. Why, one may ask, is the *partial* specific volume ($\bar{v}$) used? Why the *solution* density instead of that of the solvent? These points are highly ambiguous from the mechanical treatment, but as we shall see later, the thermodynamics of irreversible processes clearly specifies that the quantities we have used are the right ones.

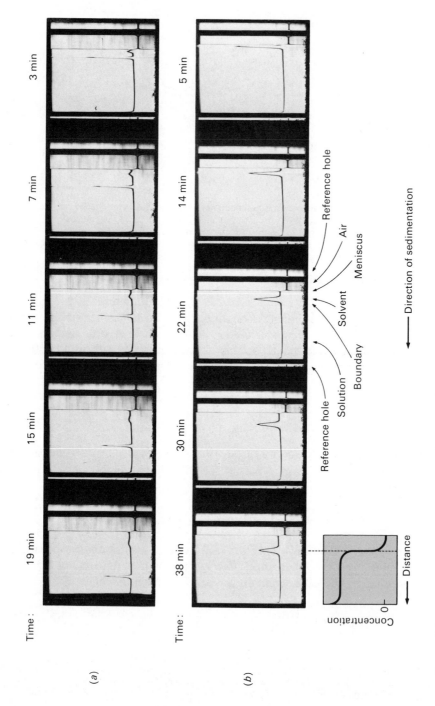

Figure 5.2 (a) Sedimentation of a hemocyanin ($M = 3,850,000$) at 42,049 rpm, 20°C. Schlieren photographs depicting the concentration gradient are given. Times after the rotor attains speeds are shown. Result: $s_{20,w} = 57.2 \times 10^{-13}$. (b) Like (a), except that these are subunits of the hemocyanin ($M \cong 380,000$, $s_{20,w} = 11 \times 10^{-13}$) sedimenting at 42,040 rpm. Note that the boundary spreads more rapidly because of the greater diffusion. The small graph shows the concentration curve corresponding to the last photograph.

The velocity divided by the centrifugal field strength $(\omega^2 r)$ is called the sedimentation coefficient, s. According to (5.3), it is proportional to the molecular weight multiplied by the *buoyancy factor* $(1 - \bar{v}\rho)$ and inversely proportional to the frictional coefficient. It is experimentally determined as the ratio of velocity to field strength. The units of s are seconds. Since values of about 10^{-13} sec are commonly encountered, the quantity, 1×10^{-13} sec is called 1 Svedberg. Svedberg units are conventionally denoted by the symbol S.

What happens when a centrifugal field is applied to a solution of large molecules? Figure 5.1 shows a diagram of an ultracentrifuge cell. Initially, we can imagine the molecules as being uniformly distributed throughout the cell. When the field is applied, all begin to move, and a region near the meniscus is cleared entirely of solute. Thus a *moving boundary* is formed; from the rate of motion of this boundary, we can calculate the sedimentation coefficient. Since $v = dr_b/dt$, we have from (5.3)

$$\frac{dr_b}{dt} = r_b \omega^2 s \qquad (5.4)$$

Integrating,

$$\ln \frac{r_b(t)}{r_b(t_0)} = \omega^2 s(t - t_0) \qquad (5.5)$$

where $r_b(t)$ is the position of the boundary at the time t. A graph of $\ln[r_b(t)/r_b(t_0)]$ versus $(t - t_0)$ will give $\omega^2 s$ and hence s.

If the diffusion coefficient were zero, the boundary would remain infinitely sharp as it traversed the cell. Such a situation is *approximated* by very large molecules at high rotor speeds. Figure 5.2(a) shows the sedimentation of a hemocyanin $(M = 3,850,000)$. However, for lower-molecular-weight materials such as the hemocyanin subunits shown in Figure 5.2(b), diffusion appreciably spreads the boundary during sedimentation. Very small molecules, with small sedimentation coefficients and large diffusion coefficients, will not form a clear boundary at all in present ultracentrifuges. The reader should note that the photographs in Figure 5.2 are obtained with a schlieren optical system (Chapter 4), which records the concentration *gradient* $(\partial c/\partial r)$; an accompanying sketch shows the concentration versus distance curve. Below the boundary there is always a *plateau* region, which in most cases will extend nearly to the bottom of the cell (see also Figure 5.3). At the bottom, of course, the solute "piles up."

Having seen roughly what happens in such a *sedimentation velocity* experiment, let us now examine the process in a somewhat more precise fashion. The sedimentation process is moving toward a state of equilibrium. From the theory of irreversible processes we would expect the flow of matter to be determined by the gradient of a potential, which will become uniform at equilibrium. But which potential? Classical thermodynamics provides the answer: For a system in a field of force, the condition for equilibrium is not that

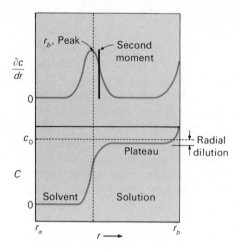

$\dfrac{\partial c}{\partial r}$

r_b, Peak

Second moment

0

c_0

Radial
dilution

Plateau

C

Solvent

Solution

0

r_a

r ⟶

r_b

Figure 5.3

An idealized picture of a boundary, showing both the C versus r and corresponding $\partial C/\partial r$ versus r curves. Sedimentation is to the right. The second moment position is to be used for exact calculation of s. It is defined by $r_b^2 = \int r^2(\partial C/\partial r)\,dr / \int (\partial C/\partial r)\,dr$, where both integrals are taken across the boundary. Since r^2 occurs in the integral in the numerator, the second moment position will be further from the center of rotation than the maximum of a symmetrical boundary.

the chemical potentials be uniform but that the *total potential* be everywhere constant. The total potential of a component is defined as the sum of the the chemical potential and the potential energy (per mole) of the substance in the field. Thus in a centrifugal field, where 1 mole of solute of molecular weight M has a potential energy of $(-\tfrac{1}{2}M\omega^2r^2)$, the total potential is $\tilde{\mu} = \mu - \tfrac{1}{2}M\omega^2r^2$, and the condition for equilibrium is

$$\left(\frac{\partial\tilde{\mu}}{\partial r}\right) = \frac{\partial\mu}{\partial r} - M\omega^2r = 0 \tag{5.6}$$

This will provide us later with a way to describe the state of *sedimentation equilibrium* in detail; for now, it allows us to write the flow equation for the solute in a two-component mixture:

$$J = -L\left(\frac{\partial\tilde{\mu}}{\partial r}\right) = -L\left(\frac{\partial\mu}{\partial r} - M\omega^2r\right) \tag{5.7}$$

Now μ is a function of T, P, and the concentration C. Therefore,

$$\left(\frac{\partial\mu}{\partial r}\right) = \left(\frac{\partial\mu}{\partial T}\right)_{P,C}\left(\frac{\partial T}{\partial r}\right) + \left(\frac{\partial\mu}{\partial P}\right)_{T,C}\left(\frac{\partial P}{\partial r}\right) + \left(\frac{\partial\mu}{\partial C}\right)_{T,P}\left(\frac{\partial C}{\partial r}\right)$$

$$\quad\quad\quad\quad\quad\quad\text{(a)}\quad\quad\quad\quad\text{(b)}\quad\quad\quad\quad\text{(c)}\quad\quad\quad\quad\text{(d)}$$

$$\left(\frac{\partial\mu}{\partial r}\right) = 0 + \bar{v}M\omega^2r\rho + \frac{RT}{C}\left(\frac{\partial C}{\partial r}\right) \tag{5.8}$$

The various terms have been evaluated as follows:

    (a) $\partial T/\partial r = 0$, since we keep T constant.
    (b) Just as $(\partial G/\partial P)_T = V$, $(\partial\mu/\partial P)_T = \bar{V}$, the partial molar volume. The partial molar volume is M times the partial specific volume.

(c) The hydrostatic pressure at a point r in the solution column is $P = P_a + \omega^2(r^2 - a^2)\rho/2$, where a is the meniscus position. This is the analog of the hydrostatic pressure in a gravitational field $g: P = P_a + g(r - a)\rho$. In both cases ρ is the *solution* density.

(d) We assume the solution to be ideal; the result follows (see Chapter 2).

Using (5.8), the flow equation (5.7) becomes

$$J = L\{\omega^2 rM(1 - \bar{v}\rho) - (RT/C)(\partial C/\partial r)\} \tag{5.9}$$

If we now express the coefficient L in terms of a frictional coefficient, $L = C/\mathcal{N}f$, we obtain

$$J = \underline{\frac{M(1 - \bar{v}\rho)}{\mathcal{N}f}}\omega^2 rC - \underline{\frac{RT}{\mathcal{N}f}}\left(\frac{\partial C}{\partial r}\right) \tag{5.10}$$

The underlined coefficients are identifiable as a sedimentation coefficient and diffusion coefficient, respectively:

$$J = s\omega^2 rC - D(\partial C/\partial r) \tag{5.11}$$

$$s = M(1 - \bar{v}\rho)/\mathcal{N}f \tag{5.12}$$

$$D = RT/\mathcal{N}f \tag{5.13}$$

Neglecting certain minor corrections, like the variation of $\bar{v}$ with pressure and the like, Equation (5.11) provides a complete description of the processes occurring in an ultracentrifuge cell. The first and second terms on the right of Equation (5.11) clearly correspond to transport by sedimentation and diffusion, respectively. When the ultracentrifuge is started, concentration is everywhere uniform $[(\partial C/\partial r) = 0]$, and only the sedimentation term is significant. However, this transport creates a boundary; the concentration gradient therein results in diffusion, with subsequent broadening. If the field were then turned off ($\omega = 0$), the boundary would stop and spread with time, as in a simple diffusion experiment.

Equation (5.11) also provides the most rigorous definition† of s; in the plateau region the flow is given by $J_p = sC_p\omega^2 r$. It can be proved that there is a point in the boundary that moves with a velocity v' such that $J_p = C_p v'$; then $s = v'/\omega^2 r$.

This point turns out to be at the second moment of the boundary gradient curve (see Figure 5.3); for reasonably sharp boundaries one can (and almost always does) assume that this point moves with the same rate as does the

---

† The necessity for a more rigorous definition than that given by (5.3) should be apparent. The discerning reader will have wondered how we define, in a broad boundary, the precise *point* to use as r_b in Equation (5.3).

point of the maximum gradient, so that only the peak of the boundary need be followed to determine s by Equation (5.5).

Equation (5.11) can be looked at as an extension of Fick's first law to include the effect of a centrifugal field. The analog of Fick's second law can also be obtained; it requires only the application of the continuity equation [see Equation (4.10)]. However, we must be careful, for the conditions in the sector-shaped ultracentrifuge cell are a little different from those used before. Since the cross section of the cell is proportional to r, the continuity equation turns out to be†

$$\left(\frac{\partial C}{\partial t}\right)_r = -\frac{1}{r}\left(\frac{\partial rJ}{\partial r}\right)_t \tag{5.14}$$

Combining (5.14) with (5.11), we obtain the partial differential equation

$$\left(\frac{\partial C}{\partial t}\right)_r = -\frac{1}{r}\left\{\frac{\partial}{\partial r}\left[s\omega^2 r^2 C - Dr\left(\frac{\partial C}{\partial r}\right)_t\right]\right\}_t \tag{5.15}$$

Solutions to this rather formidable equation have been obtained for a number of specific situations; suffice it to say here that they predict the kind of behavior illustrated in Figure 5.2 and 5.3. One initially perplexing observation can easily be explained by Equation (5.15). This is that the concentration difference across a sedimenting boundary falls off with time, rather than remaining constant as in a free diffusion experiment (see Fig. 5.3). If we apply Equation (5.15) to the plateau region, where $\partial C/\partial r = 0$, we obtain

$$\left(\frac{\partial C_p}{\partial t}\right)_r = -2s\omega^2 C_p \tag{5.16}$$

which can be integrated from $t = 0$ to give

$$C_p = C^0 e^{-2s\omega^2 t} \tag{5.17}$$

This *radial dilution effect* can be explained physically on the grounds that the flow out of any lamina in the plateau region is greater than the flow in, since both A and the field increase with r.

The sedimentation coefficient of a macromolecule provides a rough estimate of molecular weight. However, from Equation (5.12) it is evident that s is also influenced by molecular shape and dimensions, for the frictional coefficient enters into the denominator. A highly elongated molecule or a random coil will have a larger value of f, and hence a lower value of s, than a compact molecule of the same weight. The ambiguity introduced by the frictional coefficient can be taken care of by combining sedimentation and diffusion

---

† The reader can easily confirm this by repeating the derivation of Equation (4.10), letting the cross section A be proportional to r.

measurements. We must recognize that both s and D will depend on solute concentration, because of molecular interactions. If s^0, D^0, and f^0 denote quantities obtained by extrapolation to zero concentration, we have, from (5.12) and (5.13),

$$s^0 = \frac{M(1 - \bar{v}\rho)}{\mathcal{N}f^0}$$

$$D^0 = \frac{RT}{\mathcal{N}f^0}$$

$$\frac{s^0}{D^0} = \frac{M(1 - \bar{v}\rho)}{RT} \tag{5.18}$$

Thus a combination of s^0 and D^0 yields M; this is one of the more common ways of determining the molecular weights of macromolecules.

TABLE 5.1 SOME SEDIMENTATION AND DIFFUSION DATA

Substance	$T(°C)$	$s^0_{T,w} \times 10^{13}$ (sec)	$D_{T,w} \times 10^7$ (cm^2/sec)	$\bar{v}_T$ (cm^3/g)	$M_{s,D}$
Lysozyme	20	1.91	11.2	0.703	14,400
Bovine serum albumin	25	5.01	6.97	0.734	66,500
Fibrinogen	20	7.9	2.02	0.706	330,000
Polymethyl methacrylate (one fraction)	20	48.5	2.25	0.795	1,420,000
Bushy stunt virus	20	132	1.15	0.74	10,700,000

[a] Data are in most cases given at the temperature at which they were determined, since the exact information to correct to 20°C is generally not available. Many more values are recorded in H. Sober (ed.), *The Handbook of Biochemistry*, Chemical Rubber Co., Cleveland, Ohio, 1968.

In Equation (5.18) it is assumed that s and D refer to the same temperature, for the frictional coefficient depends on the viscosity, which is very temperature-sensitive. Usually, both s^0 and D^0 are corrected to some standard condition. If the actual measurements were made in a buffer solution at temperature T, the values in *water at* 20°C would be

$$s_{20,w} = \frac{(1 - \bar{v}\rho)_{20,w}}{(1 - \bar{v}\rho)_{T,b}} \frac{\eta_{T,b}}{\eta_{20,w}} s_{T,b} \tag{5.19}$$

$$D_{20,w} = \frac{293.1}{T} \frac{\eta_{T,b}}{\eta_{20,w}} D_{T,b} \tag{5.20}$$

These corrections depend on the assumption that f is accurately proportional to the solvent viscosity η; if the conditions of the experiment are not too far from water at 20°C, Equations (5.19) and (5.20) are probably valid.†

† It should be noted that Equations (5.19) and (5.10) correct only for variations of η, $\bar{v}$, and ρ with temperature and not for variations in macromolecular conformation per se. For example, if a particular molecule were compactly folded at 20°C and unfolded at 40°C, application of these equations to data obtained at 40°C would give the hypothetical s and D that the unfolded molecule would have at 20°C.

A knowledge of the sedimentation coefficient *and* the molecular weight (obtained by the s, D method or some other) allows a direct calculation of the frictional coefficient. This can provide limited information about molecular shape. If we assume that the volume of the solute molecule is given by $M\bar{v}/\mathcal{N}$, then we can calculate the radius the molecule would have were it spherical. Stokes' law [Equation (4.2)] then predicts the minimum value of the frictional coefficient (f_0) this molecule could have. If we can believe the above calculation and are willing to guess a general shape (prolate ellipsoid, oblate ellipsoid, rod, and so forth), the equations given in Table 4.1 will yield an axial ratio. But such a calculation is obviously beset with traps. Suppose the molecule is spherical but hydrated; then the frictional coefficient may be larger than f_0 just because we have calculated too small a volume by neglecting this hydration. The data appear to be really useful only when supplementary information (from viscosity measurements, for example; see Chapter 7) is available or in cases where f/f_0 is so great that we cannot reasonably attribute it to hydration. The subject will be raised again in Chapter 7.

So far. we have only alluded to the concentration dependence of s. Any interaction between the sedimenting molecules will alter the sedimentation behavior. In the most usual case the molecules may be thought of as interfering with one another, so as to make the frictional coefficient increase with concentration. If

$$f = f^0(1 + kC + \cdots) \tag{5.21}$$

where f^0 is the value of f at $C = 0$, then

$$s = \frac{s^0}{1 + kC} \tag{5.22}$$

and s will decrease with increasing C. As might be expected, this effect is most pronounced with highly extended macromolecules. Figure 5.4 shows representative data for serum albumin (a compact globular protein) and a small RNA. While both have nearly the same s^0, the concentration dependence for

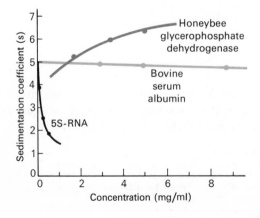

Figure 5.4

Concentration dependence of the sedimentation coefficient. Bovine serum albumin (BSA) and the ribosomal nucleic acid (5-S RNA) exhibit behavior typical of compact and extended macromolecules, respectively. The behavior of honeybee glycerophosphate dehydrogenase is that to be expected for a reversibly associating substance. Data from R. L. Baldwin (see references), D. G. Comb, and Zehavi-Willmer, *J. Mol. Biol.*, **23**, 441 (1967), and R. R. Marquardt and R. W. Brosemer, *Bioch. Biophys. Acta*, **128**, 454 (1966).

the RNA is much greater. This is one reason absorption optical systems have come to be so widely used in studying the sedimentation of nucleic acids.

Also shown in Figure 5.4 are data from a rather different kind of system, in which *s increases* with increasing C. In all cases in which such behavior has been observed, it appears to result from a rapid monomer-polymer equilibrium. Under some conditions, such systems will yield only a single boundary, with sedimentation coefficient dependent on the proportions of monomer and polymer. Since high concentrations favor polymer formation, the *s* value increases with C.

If there are a number of noninteracting solute components, a sedimentation velocity experiment *may* lead to their resolution. Such a situation is depicted in Figure 5.5(*a*). If there is very little solute-solute interaction, the relative amounts of the components, as well as their individual sedimentation coefficients, can be determined. However, if the sedimentation coefficients are mutually concentration-dependent, difficulties arise in the analysis. An example is shown in Figure 5.5(*b*). Here, each component sediments more slowly in the presence

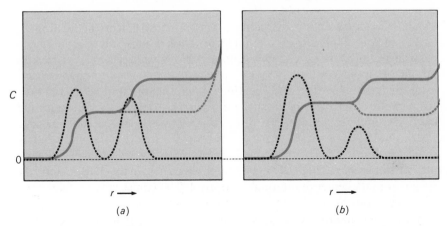

Figure 5.5   Sedimentation of a mixture of two macromolecules without (*a*) and with (*b*) the Johnston-Ogston effect. The broken curves show the gradients that would be observed.

of the other than it would alone. For the fast component, which is always passing through a solution of the slow component, a reduced *s* value is observed. The analysis of the composition is also more complicated. Since the slow component is retarded below the fast boundary but not above, it will tend to pile up behind the fast boundary. Most optical techniques yield only the total concentration change across each boundary, so that the composition of the system in Figure 5.5(*b*) would be misread. One would obtain a value too high for the amount of slow component and a value too low for the amount of fast component. This "Johnston–Ogston" effect can lead to very large errors in highly concentration-dependent systems.

Last, a word about homogeneity. While the observation of more than one sedimenting boundary is proof of heterogeneity, the converse is not necessarily true. Either high concentration dependence or rapid diffusion can make it impossible to resolve adjacent boundaries in the ultracentrifuge. Normally, sedimentation coefficients decrease with increasing concentration, so that molecules that diffuse to the low-concentration side of a boundary will tend to "catch up," giving an artificially sharp boundary, which can conceal heterogeneity.† Or, if the diffusion rate is too great as compared to the rate of sedimentation, two gradient curves may so overlap as to yield a broad boundary with a single maximum, difficult to distinguish from that of a single component. Thus the frequently encountered statement that the "sample was found to be homogeneous in the ultracentrifuge" should be taken with a grain of salt unless rather rigorous tests have been applied to the data or unless the sample has been examined over a very wide range of experimental conditions.

5.2 APPARATUS AND PROCEDURES FOR SEDIMENTATION STUDIES

Quantitative sedimentation experiments require instruments that can provide high, controlled rotor speeds at uniform temperatures. Most modern ultracentrifuges utilize an electric motor drive, with either mechanical or electronic automatic speed control. In addition, the rotor is usually operated in an evacuated chamber to reduce heating by air friction, and refrigeration and temperature control is generally provided. Two types of ultracentrifuges, *analytical* and *preparative*, are distinguished by whether or not an optical system for the observation of the cell is provided (see Chapter 4 for details of optical systems). Some commercially available analytical ultracentrifuges can attain rotor speeds as high as 74,000 rpm, giving a field at the cell position (about 6.8 cm from the center of rotation) of nearly 420,000 times gravity. Illustrations of such equipment are shown in Figures 5.6(*a*) and 5.6(*b*).

Since sedimentation velocity experiments are so widely used in biochemistry, a few words about proper procedure are in order. A careful study will include measurements at several concentrations; these should extend to as low a concentration as the optical system and substance will allow. For biopolymers, a buffer solution, usually with ionic strength at least 0.1, should be used. This is because charged polymers may be retarded in sedimentation by an effect analogous to the Donnan effect in osmometry (see Chapter 2). Temperatures should be controlled precisely, and the rotor speed should be actually measured during the experiment. The photographic plates that record the schlieren image should be measured with a microcomparator. For the most accurate work, especially with broad boundaries, the second moment should be used

---

† Extreme examples of concentration-dependence sharpening are found with nucleic acids. Because of the great dependence of *s* on *C*, even very heterogeneous samples will give razor-sharp boundaries at high concentrations.

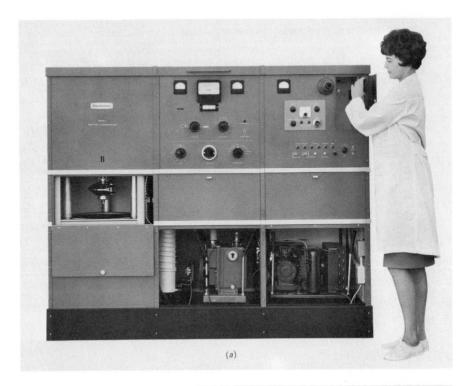

Figure 5.6
(a) A modern analytical ultra-centrifuge. Courtesy Beckman Instruments Co. (b) Rotor and typical cell for an analytical ultracentrigue. Courtesy Beckman Instruments Co.

for the boundary; for routine studies the maximum in the gradient can be employed.

With proper attention to such details, sedimentation coefficients can be determined with precision of better than 0.5 percent. Since D can also be quite accurately measured, precise molecular weights should be obtainable from Equation (5.18). Actually, the quantity hardest to obtain with high accuracy is the partial specific volume. Since $\bar{v} \simeq 0.7$ for most proteins, an error of 1 percent in $\bar{v}$ gives an error of about 3 percent in $(1 - \bar{v}\rho)$ and, consequently, in M.

5.3 SEDIMENTATION EQUILIBRIUM

Now let us examine that state of equilibrium toward which a sedimentation experiment is moving. We shall be thinking of an experimen in which the rotor speed is not so great as in a velocity experiment; we do not want all of the solute to be packed in the bottom of the cell. To obtain equations describing the gradient that will exist at equilibrium, we might start with the general equilibrium condition [Equation (5.6)] without every introducing the concept of flow. However, to save a lot of duplication in the derivations, let us take the following point of view: When equilibrium is finally established, flow must vanish at all points in the ultracentrifuge cell. There will still be the random motion of molecules, but no net transport of matter will henceforth occur, so long as the ultracentrifuge continues to operate at the same rotor speed and temperature. Thus we must set $J = 0$ in Equation (5.9):

$$L\left\{\omega^2 rM(1 - \bar{v}\rho) - \frac{RT}{C}\frac{dC}{dr}\right\} = 0 \tag{5.9}$$

$$\frac{1}{C}\frac{dC}{dr} = \frac{\omega^2 rM(1 - \bar{v}\rho)}{RT} \tag{5.23}$$

We have replaced the partial derivative of (5.9) with a total derivative, since C is no longer a function of time but only of r.

Equation (5.23) decribes the concentration gradient at equilibrium for a single solute component in an ideal two-component solution. By integrating it between the meniscus (a) and some point r, we can see that C depends on r in an exponential fashion:

$$C(r) = C(a)e^{\omega^2 M(1 - \bar{v}\rho)(r^2 - a^2)/2RT}$$

or

$$\ln\frac{C(r)}{C(a)} = \frac{\omega^2 M(1 - \bar{v}\rho)(r^2 - a^2)}{2RT} \tag{5.24}$$

A graph of $\ln C(r)$ versus r^2 will yield the molecular weight. A photograph, obtained with an interferometric optical system (see Chapter 4), of a system at sedimentation equilibrium is shown in Figure 5.7(a). Each interference fringe can be thought of as tracing the variation of C with r in the cell. Note, however, that neither the interference fringes nor integration of a schlieren pattern will give the absolute concentration; rather, each gives the difference in concentration between a point r and some reference point, such as the meniscus. To obtain the absolute concentrations, two methods are used. First, if the ultracentrifuge is operated at a rotor speed so high that the meniscus concentration is negligible, the absolute concentration at any point is equal to the concentration difference

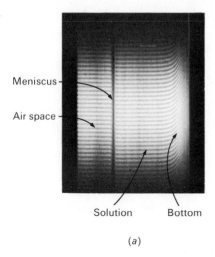

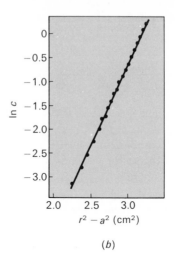

(a)   (b)

Figure 5.7   A sedimentation equilibrium experiment with 30-S ribosomes from *Escherichia coli*. The interferogram is shown in (a), while (b) depicts a graph according to Equation (5.24). Conditions: $C^0 = 0.35$ mg/ml; $T = 4°C$.

between that point and the meniscus. This is the basis of the so-called Yphantis method or high-speed method. This has been done in the example in Figure 5.7(a). An alternative is to utilize the conservation of mass [see Equation (5.28)] to calculate the concentration at the meniscus. The reader can find how this is done by writing out the expression for the experimentally determinable integral

$$\int_a^b [C(r) - C(a)]r \, dr,$$

and using (5.28). The corresponding graph according to Equation (5.24) is shown in Figure 5.7(b). With such data, only $\bar{v}$, ρ, T, and ω need also be known to calculate M. Note that there is no frictional coefficient involved here; such a quantity has no place in the description of a state of thermodynamic equilibrium.

Equation (5.23) can easily be extended to more general cases; for example, for a nonideal solution, it becomes

$$\frac{1}{C}\frac{dC}{dr} = \frac{\omega^2(1 - \bar{v}\rho)Mr}{RT\{1 + C(\partial \ln y/\partial C)\}} \tag{5.25}$$

[It would be a good idea if the reader were to do Problem 10 at this point; deriving (5.25) will review the derivation of (5.23).]

If the solution is ideal but the solute is heterogeneous, an equation of the form of (5.23) can be written for each solute component, *i*. We can then show how sedimentation equilibrium can yield *average* molecular weights for such solutes. The equations for the individual components can be integrated in a fashion different than was used for (5.24):

$$\int_a^b \frac{dC_i}{dr}\,dr = \frac{\omega^2 M_i(1-\bar{v}\rho)}{RT}\int_a^b C_i r\,dr \tag{5.26}$$

The integral on the left is trivial; it is just $[C_i(b) - C_i(a)]$. The one on the right looks more formidable, but it is easily found from the fact that the total mass of component i in the sector-shaped cell is constant during the experiment. If the sector angle of the cell is ϕ and its thickness is h, we have (see Figure 5.1 again)

Mass of i in cell at equilibrium = mass of i in cell at start

$$\underbrace{\int_a^b C_i(r)h\phi r\,dr}_{\substack{\text{volume element}\\ \text{at } r}} = \int_a^b \underbrace{C_i^0}_{\substack{\text{original concentration}\\ \text{(uniform in cell at } t=0)}} h\phi r\,dr = C_i^0\int_a^b h\phi r\,dr \tag{5.27}$$

Or, since h and ϕ are constants,

$$\int_a^b C_i(r)r\,dr = C_i^0\int_a^b r\,dr = C_i^0\frac{b^2 - a^2}{2} \tag{5.28}$$

We may now put this into (5.24) and sum over all q solute species:

$$\sum_{i=1}^q [C_i(b) - C_i(a)] = \frac{\omega^2(1-\bar{v}\rho)}{RT}\frac{b^2 - a^2}{2}\sum_{i=1}^q C_i^0 M_i \tag{5.29}$$

We have assumed for convenience that all $\bar{v}_i = \bar{v}$, a constant value. Recalling that the sums of the solute concentrations represent the total solute concentration,

$$\frac{C(b) - C(a)}{C^0} = \frac{\omega^2(1-\bar{v}\rho)}{RT}\frac{b^2 - a^2}{2}\frac{\displaystyle\sum_{i=1}^q C_i^0 M_i}{\displaystyle\sum_{i=1}^q C_i^0} \tag{5.30}$$

The quantity $\sum C_i^0 M_i / \sum C_i^0$ has been identified earlier (see Chapter 2) as the weight-average molecular weight. Equation (5.30) says that it can be measured simply by determining the concentration difference between the ends of the solution column at equilibrium and dividing by the initial concentration. The same data, if treated in another way, will yield the *Z-average molecular weight*, $M_z = \sum C_i^0 M_i^2 / \sum C_i^0 M_i$, and under some circumstances the number-average molecular weight as well.† Thus a sedimentation equilibrium experiment pro-

---

† There are a variety of ways to analyze sedimentation equilibrium data. The interested reader is referred to any of the review articles cited at the end of this chapter. See H. Fujita's book (1962) and Problem 8.

vides information on heterogeneity; conversely, identity of the various averages is an excellent proof of homogeneity.

It is probably safe to say that sedimentation equilibrium is the most accurate physical method now available for the determination of macromolecular weights. The method is not especially time-consuming (about 24 hr for moderate molecular weights) if short solution columns (ca. 3 mm) are used. Table 5.2 gives some values that have been obtained for proteins with known amino acid sequence (and, consequently, precisely known molecular weight). The data for sucrose and *E. coli* ribosomes are included to show the range of the method.

TABLE 5.2   SOME SEDIMENTATION EQUILIBRIUM RESULTS

Substance	Molecular Weight from Chemical Formula	Molecular Weight from Sedimentation Equilibrium
Sucrose	342.3	341.5
Ribonuclease	13,683	13,740
Lysozyme	14,305	14,500
Chymotrypsinogen *A*	25,767	25,670
30-S *E. coli* ribosomes	—	900,000

5.4   ARCHIBALD METHOD

Our proof of the equations for sedimentation equilibrium depended on the fact that as equilibrium is reached, flow vanishes at every point in the cell. But throughout any sedimentation experiment, there are two points in the cell where the flow is zero at *all* times—the meniscus and the cell bottom. This statement rests on the fact that matter cannot be pushed through either end of the solution column. Thus Equations (5.23) and (5.25) must apply *at these two points* at any time from the beginning of the experiment. The values of dC/dr and C are, of course, those pertaining to the cell bottom or meniscus. For a homogeneous material, the molecular weights calculated at these two points should be the same. For a heterogeneous solute the situation is more complicated, since a redistribution of solute components occurs as sedimentation proceeds. The apparent molecular weight at the meniscus will decrease with time, while that at the cell bottom will increase. Each should approach the weight-average molecular weight of the solute if extrapolated back to zero time. This effect results, of course, from the fact that higher-molecular-weight material is being selectively removed from the meniscus and accumulates at the bottom.

The Archibald method, named for the scientist who pointed out the possibility of such measurements, has the advantage of rapidity over the conventional sedimentation equilibrium experiment. However, since it depends on data from the extremities of the solution column, where precision of measurement is poorest, it is usually inferior in accuracy.

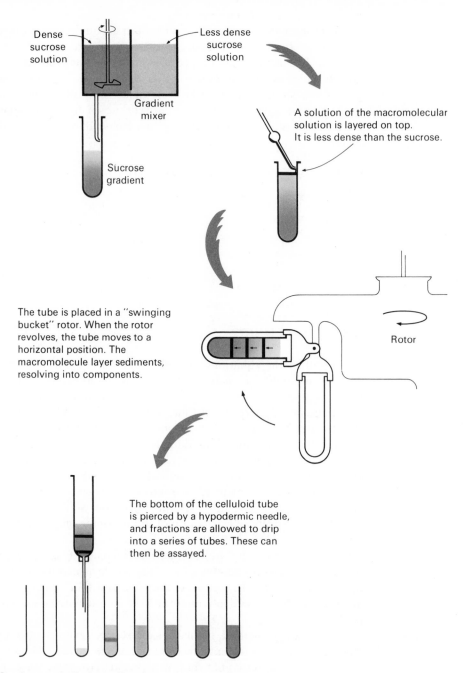

Dense sucrose solution

Less dense sucrose solution

Gradient mixer

Sucrose gradient

A solution of the macromolecular solution is layered on top. It is less dense than the sucrose.

The tube is placed in a "swinging bucket" rotor. When the rotor revolves, the tube moves to a horizontal position. The macromolecule layer sediments, resolving into components.

Rotor

The bottom of the celluloid tube is pierced by a hypodermic needle, and fractions are allowed to drip into a series of tubes. These can then be assayed.

Figure 5.8   Sucrose gradient centrifugation.

5.5 DENSITY GRADIENT SEDIMENTATION

If sedimentation of a macromolecule is carried out in the presence of a density gradient that has been produced by a gradient in the concentration of some "inert" substance, the resolving power of the ultracentrifuge can be greatly enhanced. There are two kinds of techniques: In the equilibrium method, the concentration gradient in the inert substance is obtained by running the ultracentrifuge until equilibrium is attained, and the macromolecular component comes to equilibrium in this gradient. Alternatively, one may produce a density gradient with a substance such as sucrose by carefully layering solutions of decreasing density and then studying the sedimentation transport of a macromolecular material through this gradient.

Let us consider the latter technique first. The sequence of operations as usually performed is shown in Figure 5.8. The sucrose gradient produced in this way is, of course, not stable; it will eventually disappear by diffusion. But this process is slow, and experiments of several hours' duration can be performed without too much change occurring in the gradient. The macromolecular solution can be layered onto such a gradient, providing that it is quite dilute and less dense than the top sucrose solution.

The advantage in resolving power of the gradient technique over the conventional sedimentation transport experiment is obvious. In the conventional method, only the most slowly sedimenting material can be recovered in pure form, whereas the density gradient method gives a *spectrum* of the mixture. The crude-appearing sampling technique actually works well and little remixing is produced. From the tube number in which a given component appears, the distance traveled can be estimated and s can be calculated. However, the variation in both density and viscosity in the sucrose gradient makes the velocity vary during sedimentation and complicates the calculations. While the inclusion of marker substances of known sedimentation coefficient increases the accuracy, the method still cannot give the precision of the conventional techniques. Density gradient centrifugation does, however, have the enormous advantage that a wide variety of assay techniques can be used to detect the various components. Materials can be specifically labeled with radioactive isotopes or discriminating enzymatic assays can be used. In this way the sedimentation of a minor component in a crude mixture can be followed, a feat impossible with the optical methods used in most analytical ultracentrifuges.† Figure 5.9 shows the sedimentation of an enzyme mixture in a sucrose gradient.

The density gradient sedimentation equilibrium experiment works on an entirely different principle. Suppose, as a specific example, that a solution

---

† Recently, a variant in the density gradient velocity experiment has been introduced that allows use of the analytical ultracentrifuge. It utilizes a special cell in which a layer of the solution is automatically layered on a salt solution as the ultracentrifuge is accelerated. Diffusion at the edge of the layer produces a density gradient that stabilizes the zone. The sedimentation of bands can then be observed with the usual optical system.

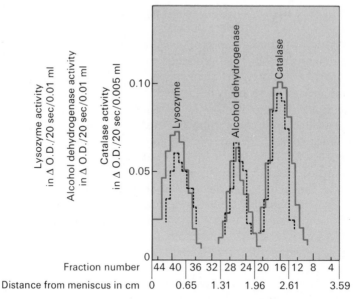

Figure 5.9 A sucrose gradient sedimentation velocity experiment with mixtures of three enzymes. The solid lines show enzyme activities in a mixture of the three enzymes in tris buffer. The dotted lines show activities in a mixture prepared using a crude bacterial extract as the solution medium. In each case, sedimentation was for 12.8 hr at 3°C, 37,700 rpm. From R. B. Martin and B. Ames, *J. Biol. Chem.*, **236**, 1372 (1961).

containing a small amount of high-molecular-weight nucleic acid and a high concentration of a dense salt such as cesium chloride is placed in an ultracentrifuge cell and spun for a long time at high speed. The salt will eventually reach sedimentation equilibrium, giving a concentration gradient like that described by Equation (5.24). (Actually, this equation will not be obeyed exactly, for the concentrated solution will be nonideal.) This will yield a density gradient in the cell. The experiment becomes interesting if the salt concentration has been so chosen that at some point (r_0) in the cell, $\rho(r_0) = 1/\bar{v}$, where $\bar{v}$ is the specific volume of the macromolecule. The quantity $(1 - \bar{v}\rho)$ will then be positive above this point and negative below it, so we would expect sedimentation of the macromolecule to converge to the point r_0. Eventually, equilibrium will be established between the tendency for the macromolecules to diffuse away from r_0 and the effect of sedimentation to drive them back.

We can use Equation (5.23) to describe the sedimentation equilibrium of the macromolecule providing we remember that ρ depends on r. For the limited region about r_0 in which we are interested, we may assume that the density gradient is linear:

$$\rho(r) = \frac{1}{\bar{v}} + (r - r_0)\frac{d\rho}{dr} \tag{5.31}$$

where the value of $d\rho/dr$ is determined by the nature of the salt and its concentration, as well as the rotor speed. From (5.23), then,

$$\frac{1}{C}\frac{dC}{dr} = \frac{\omega^2 rM}{RT}\left\{1 - \bar{v}\left[\frac{1}{\bar{v}} + (r - r_0)\frac{d\rho}{dr}\right]\right\}$$

$$= -\frac{\omega^2 rM}{RT}\bar{v}(r - r_0)\frac{d\rho}{dr} \tag{5.32}$$

Recognizing that $r \simeq r_0$ and $(r - r_0)\,dr = (1/2)d(r - r_0)^2$, this becomes

$$d \ln C \simeq -\frac{\omega^2 r_0 M}{2RT}\bar{v}\frac{d\rho}{dr}d(r - r_0)^2 \tag{5.33}$$

or, upon integration,

$$C(r) = C(r_0)\exp\left[-\frac{\omega^2 r_0 M\bar{v}}{2RT}\frac{d\rho}{dr}(r - r_0)^2\right] \tag{5.34}$$

Thus at equilibrium the macromolecular component will be distributed in a Gaussian band about r_0. The breadth of this band will depend on $M^{-1/2}$; thus a substance of infinite M would be packed into an infinitely thin band. The measurement of molecular weight is complicated by the disastrous effects of even minor density heterogeneity. Obviously, a narrow distribution of densities will produce a broadened band that will be difficult to distinguish from the Gaussian band for a lower-weight substance.

While the method can be used for molecular-weight determination, its greatest utility lies in the sensitivity of the band position to solute density. Figure 5.10 shows the clear resolution of nucleic acids differing only in that one contains ^{14}N and the other ^{15}N. The reader should consult Wold, in this series, to appreciate the important role that this method has played in modern molecular biology.

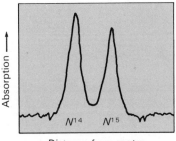

Figure 5.10

Density gradient sedimentation equilibrium of a mixture of DNA's from bacteria grown on ^{15}N and ^{14}N media. Note that separation of the bands is excellent, even though the densities differ only from the isotope substitution. From M. Meselson, F. W. Stahl, and J. Vinograd, *Proc. Natl. Acad. Sci. US*, **43**, 581 (1957).

Our treatment of the theory has been somewhat unsophisticated, since we have neglected solute-solute and solute-solvent interactions. A more exact discussion, in which the nonideality is considered, yields the important difference that in this case the macromolecular weight and specific volume refer to the *solvated* molecule. We have erred in treating the system as one containing only two components. Since a high concentration of the salt is required, a full-fledged three-component treatment is needed for accurate interpretation.

PROBLEMS

1. In a study of the sedimentation of bovine serum albumin, the following data were obtained:

t (sec)	r_b/r_0
700	1.0129
3,580	1.0679
4,540	1.0871
5,020	1.0965

The rotor speed was 59,780 rpm. What is the sedimentation coefficient, s?

2. The best data for bovine serum albumin at 25.00°C are

$$s^0 = 5.01 \times 10^{-13} \text{ sec}$$

$$D^0 = 6.97 \times 10^{-7} \text{ cm}^2/\text{sec}$$

$$\bar{v} = 0.734 \text{ cm}^3/\text{g}, \quad \rho = 1.0024 \text{ g/cm}^3$$

Calculate M, f/f_0, and the axial ratio, assuming the molecule is an unhydrated prolate ellipsoid.

3. A sedimentation velocity experiment was performed with subunits of aldolase at 3°C, with a rotor speed of 51,970 rpm. Data for the boundary position versus time are as follows. The time of the first photograph has been taken as zero.

r_b (cm)	$t - t_0$ (sec)
6.2438	0
6.3039	3,360
6.4061	7,200
6.4920	11,040
6.5898	14,840
6.6777	18,720

Calculate the sedimentation coefficient. Correct to $s_{20,w}$, assuming $\bar{v} = 0.73$ at 3°C and 0.74 at 20°C; other data can be found in standard handbooks. Assume solvent is water.

*4. For a heterogeneous material, we can define a weight-average sedimentation coefficient and a weight-average diffusion coefficient by

$$\bar{s}_w = \frac{\sum s_i C_i}{\sum C_i}; \qquad \bar{D}_w = \frac{\sum D_i C_i}{\sum C_i}$$

Show that for a polymer for which the frictional coefficient is proportional to some power (b) of the molecular weight ($f_i = K M_i^b$), the M calculated from s_w and D_w is *not* in general a weight-average molecular weight. What average will be obtained if $b = 1$? If $b = 0$?

5. Bacterial flagellin is a rodlike protein molecule of molecular weight 20,000. It can be caused to dimerize by a slight change in pH or ionic strength. The sedimentation coefficient for the monomer is $1.74s$, and for the dimer, $2.36s$. Which of two probable (side-to-side or end-to-end) aggregation modes is supported by the data? Prove your answer.

6. For a homologous series of random-coil molecules, one might expect the frictional coefficient to be proportional to the radius of gyration. We observe that such series (that is, polymers of a given monomer) obey equations such as

$$s = K M^a$$

What would you predict the value of a to be? For DNA, a is found to be about 0.33. How can this result be explained?

*7. For ribonuclease,

$$M = 13,683, \qquad \bar{v} = 0.695 \text{ cm}^3/\text{g}$$

In a sedimentation equilibrium experiment with a solution of ribonuclease containing 0.780 g/100 ml in a very dilute buffer at 25.00°C and a rotor speed of 18,600 rpm, calculate the equilibrium concentrations at the cell bottom and meniscus if b and $a = 7.013$ and 6.750 cm, respectively. [*Hint:* Calculate $C(b) - C(a)$ and $C(b)/C(a)$; combine.] What is the *exact* concentration half-way down the solution column?

*8. Show that the Z-average molecular weight can be obtained from the sedimentation of an ideal heterogeneous solute by the equation

$$\frac{(1/b)(dC/dr)_b - (1/a)(dC/dr)_a}{C_b - C_a} = \frac{M_z(1 - \bar{v}\rho)\omega^2}{RT}$$

where $(dC/dr)_b$ and $(dC/dr)_a$ are the concentration gradients at the bottom and meniscus, respectively. If the schlieren deflection $Z = K(dC/dr)$ (K is a constant), show that this gives

$$\frac{(Z_b/b) - (Z_a/a)}{\int_a^b Z \, dr} = \frac{M_z(1 - \bar{v}\rho)\omega^2}{RT}$$

9. A sedimentation equilibrium experiment on rabbit muscle aldolase subunits was carried out at pH 2.9, $T = 2.0°C$, rotor speed $= 12,500$ rpm. At this temperature, $\bar{v} = 0.726$, and solution density $= 1.009$. The data are

	r (cm)	Z (cm)
Meniscus	6.842	—
	6.885	0.0950
	6.928	0.1252
	6.972	0.1734
	7.015	0.2456
	7.059	0.3256
Bottom	7.069	—

Values of Z, the schlieren wire image deflection, cannot be read directly at the meniscus and bottom and must be obtained at these points by extrapolation. Z is proportional to dC/dr, and the units of Z will cancel in your calculations. Obtain the Z-average molecular weight, using the result of Problem 8. This answer turned out to be a poor approximation to the molecular weight of the subunit. Suggest why.

10. Starting with the condition for equilibrium in a centrifugal field, $d\tilde{\mu}/dr = 0$, Derive Equation (5.25) for a two-component nonideal mixture.

11. It is believed that protein synthesis occurs on polysomes, which are aggregates of ribosomes. Devise a sucrose gradient experiment that will test this.

12. The buoyant density of *E. coli* DNA in CsCl is about 1.712 g/cm^3, whereas that for salmon sperm DNA is about 1.705 g/cm^3. What would be the peak-to-peak separation and half-widths of the peaks in a density gradient equilibrium experiment on a mixture of these two DNA's if each had a molecular weight of 2×10^6? You may assume that the rotor speed is 45,000 rpm and that the density gradient is 0.1 g/cm^4. Would the resolution be improved by dropping the speed to 30,000 rpm? You may assume that $d\rho/dr$ is proportional to ω^2, and that $\rho = 1.710$ g/cm^3 at $r = 6.5$ cm. Assume $T = 20°C$.

REFERENCES

General

Schachman, H. K.: *Ultracentrifugation in Biochemistry*, Academic Press, Inc., New York, 1959. Much on technique; a good, overall resume for biochemists.

Svedberg, T., and K. O. Pedersen: *The Ultracentrifuge*, Oxford University Press, New York, 1940. The classic reference, somewhat dated.

Fujita, H.: *Mathematical Theory of Sedimentation Analysis*, Academic Press, Inc., New York, 1962. Described by its title; the ultimate reference for questions of theory.

Williams, J. W., K. E. Van Holde, R. L. Baldwin, and H. Fujita: *Chem. Rev.*, **58**, 715 (1958). A general review, emphasizing theory.

Specific

Baldwin, R. L.: *Biochem. J.*, **65**, 503 (1957). An example of careful *s* measurement.

Meselson, M. and F. W. Stahl: *Proc. Natl. Acad. Sci. US*, **44**, 671 (1958). An elegant use of the density gradient method.

Van Holde, K. E. and R. L. Baldwin: *J. Phys. Chem.*, **62**, 734 (1958). The low-speed equilibrium method.

Yphantis, D. A.: *Biochemistry*, **3**, 297 (1964). The high-speed equilibrium method.

ELECTRIC FIELDS

SIX

The great majority of the polymers of biological interest are electrically charged. Like low-molecular-weight electrolytes, polyelectrolytes are somewhat arbitrarily classified as "strong" or "weak" depending on the ionization constants of the acidic or basic groups. They may be strong polyacids, such as the nucleic acids; weak polybases, such as poly-*l*-lysine; or polyampholytes, such as the proteins. As shown in Chapter 2, this polyelectrolyte character has a considerable influence on the solution behavior of these substances. It is not surprising, then, that there has been a great deal of interest in methods for the determination of the charge carried by biopolymers. Conversely, the fact that otherwise similar molecules may sometimes differ in charge has been widely used as the basis for electrical separation methods. Finally, the fact that the distribution of charge on a macromolecule may be asymmetric (even when the net charge is zero) has been utilized to orient macromolecules in electric fields.

It should be recalled that there are two basic ways in which a molecule can respond to an applied electrical field. If it carries a net charge, it will *migrate* in the field; this is the basis of all studies of electrophoresis. In addition, the molecule may either have an asymmetric distribution of charge or be polarizable. In such case, an electric field will produce some degree of *orientation*. It is the purpose of this chapter to examine these two kinds of responses to electrical fields and to show how they can be used.

6.1 TRANSPORT IN AN ELECTRIC FIELD: ELECTROPHORESIS

The transport of particles by an electrical field is termed electrophoresis. Formally, the process is very like sedimentation, for in both cases a field, which is the gradient of a potential, produces the transport of matter. However, since electrophoresis will depend on the charge on a macromolecule rather than its mass, it provides an additional "handle" for the analysis and separation of mixtures.

We might hope to use the theory of irreversible processes to develop a very general theory of electrophoretic transport (and, in fact, some progress has been made in this direction) but certain complications in the analysis of real systems make a mechanistic point of view more fruitful for our purposes. These difficulties arise mainly because electrophoresis is almost always carried out in aqueous solutions containing buffer ions and salts in addition to the macromolecule of interest; at the very least there must be present the counterions that have dissociated from the macromolecule to provide its charge. This means that we study the molecule not alone but in the presence of many other charged particles, and these will both influence the local field and interact with the macromolecule so as to make analysis difficult. To begin, then, let us consider the unreal simplified situation, an isolated charged particle in a nonconducting medium [Figure 6.1(a)]. We may write the basic equation for

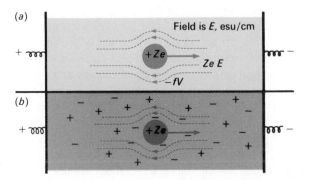

Figure 6.1   (a) An idealized model in which a particle of charge Ze is placed in an electric field in a nonconducting solvent. (b) A more realistic model, in which a charged macromolecule is subjected to an electric field in an aqueous salt solution. The small ions form an ion atmosphere around the macromolecule in which ions of opposite charge to that of the macromolecule predominate. This ion atmosphere is distorted by the field and by the motion of the macromolecule.

electrophoresis in exactly the same fashion as was used in sedimentation; we assume that the force on the particle, when suddenly applied, causes an immediate acceleration to a velocity at which the viscous resistance of the medium just balances the driving force. The force (in dynes) experienced by a particle in an electrical field is given by Coulomb's law:

$$F = zeE \tag{6.1}$$

where z is the number of electron units of charge ($e = 4.8 \times 10^{-10}$ esu)† and E is the field in electrostatic units of potential per centimeter. Then the velocity (v) is given by

$$fv = zeE \tag{6.2}$$

where f is a frictional factor, which, in this case at least, we can identify with the factor that would be determined by a sedimentation or diffusion experiment with the same particle. Equation (6.2) can be rearranged to define an *electrophoretic* mobility, a quantity very analogous to the sedimentation coefficient:

$$U = \frac{v}{E} = \frac{ze}{f} \tag{6.3}$$

If the particle happens to be spherical, Stokes' law applies and one may write

$$U = \frac{ze}{6\pi\eta R} \tag{6.4}$$

where R is the particle radius and η the solvent viscosity. In principle this equation could be generalized to take into account hydration and deviations from spherical shape, as was done in the cases of sedimentation and diffusion. However, in dealing with real macroions in aqueous solution, there are more formidable complications to consider first. In such an environment, the macromolecule is surrounded by an *ion atmosphere*, a region in which there will be a statistical preference for ions of opposite sign [see Figure 6.1(b)]. This means that the field experienced by the macromolecule will differ appreciably from that which we measure as being imposed by the electrodes. To a degree, correction for this effect can be made if one uses the Debye–Hückel theory to calculate the properties of this ion atmosphere. A rough approximation gives

$$U = \frac{ze}{6\pi\eta R} \frac{X(\kappa R)}{1 + \kappa R} \tag{6.5}$$

where

$$\kappa = \left(\frac{8\pi \mathcal{N} e^2}{1,000 DkT} \right)^{1/2} I^{1/2} \tag{6.6}$$

---

† It may be wondered why electrostatic units are used in these calculations. It is because calculation of force, energy, and other mechanical quantities in centimeter-gram-second units follows directly from the use of electrostatic units for electrical quantities. The reader is reminded that 1 V equals 1/300 esu of potential, and 1 C equals 3×10^9 esu of charge.

and D is the dielectric constant of the medium and I the ionic strength. The quantity κ is the "reciprocal ion-atmosphere radius" from Debye–Hückel theory; it depends on the ionic strength of the medium. For large ionic strengths, the atmosphere is shrunk tightly around the macroion; if I is small, the atmosphere is diffused and extended. The function $X(\kappa R)$ is called Henry's function; it varys between 1.0 and 1.5 as κR goes from zero to infinity (see Figure 6.2).

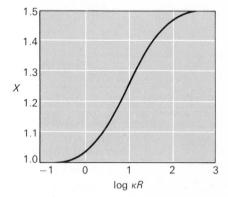

Figure 6.2

Henry's function. R is the macromolecular radius and κ the Debye–Hückel parameter.

[At this point it would be useful to work out Problem 3, so as to obtain a feeling for the quantities involved and to see the effect of the correction terms in Equation (6.5).] However, even Equation (6.5) does not provide an accurate description of the electrophoretic mobilities of real macroions. This is because no correction has been made for the fact that the ion atmosphere is distorted both by the field and by the motion of the particle through the medium [see Figure 6.1(b)]. More complex expressions have been derived, but the situation is still far from satisfactory. The net result of all of these complications is that electrophoresis, while providing an excellent method for separation of macromolecules, has proved less useful than methods such as sedimentation for obtaining information about macromolecular structure.† It is exceedingly difficult, for example, even to obtain a reliable value for the macromolecular charge, z. Inaccurate as it may be, Equation (6.5) predicts one point of interest; at the pH at which a protein or other polyampholyte has zero charge (the isoelectric point) the mobility should be zero. At lower pH values a protein will be positively charged and move toward the negative electrode, and at higher pH transport will be in the opposite direction. Thus the measurement of isoelectric points can be made simply and unambiguously by means of electrophoresis. An example is shown in Figure 6.3. In this case, an attempt has been made to calculate the mobility of a protein at pH values different from the

† The difficulty of resolving these complications has led to a decline in interest in electrophoresis as a tool for the study of macromolecular properties. Most of the emphasis in recent years has centered on the development of rapid and sensitive *separation* methods, many of them involving solid supporting media. See Table 6.1.

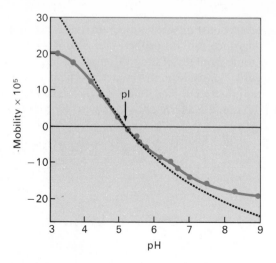

Figure 6.3

Mobility versus pH for the protein β-lactoglobulin. The points and the solid curve show experimental results, while the broken curve shows the mobility predicted by Equation (6.5), using titration data for the charge. Taken from R. K. Cannan, A. H. Palmer, and A. C. Kibrick, *J. Biol. Chem.* **142**, 803 (1942).

isoelectric point. Comparison of the result with the experimentally determined electrophoretic mobility demonstrates a considerable difference. Some of this may result from the binding of anions or cations in the buffer solution.

A number of techniques are available for electrophoresis; some are best for preparative separations, whereas others allow the precise control of conditions necessary for mobility measurement. The most accurate method in the latter respect is the *moving boundary* method. The cell that is used is depicted in Figure 6.4; it is essentially a U-tube, with sliding joints that facilitate the formation of sharp boundaries. The electrodes are contained in side arms, so that products of the electolysis process do not fall into the boundary region. The boundaries between buffer and protein solution are formed as shown in the figure and then gently displaced into the straight limbs of the cell. These limbs

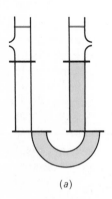

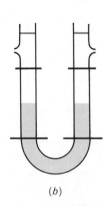

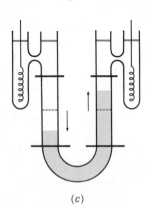

(a)   (b)   (c)

Figure 6.4   Formation of boundaries in a Tiselius electrophoresis apparatus. The U-tube is filled, as in (a), with solution (shaded) on one side and solvent (unshaded) on the other. In (b) the sliding partitions have opened, and the two boundaries have formed. In (c) electrodes have been connected, and electrophoresis has begun.

are provided with flat windows in planes parallel to the paper. Through these windows the boundaries can be observed, using either a schlieren or interferometric optical system. The same apparatus, without the electrode compartments can be used for diffusion studies (see Chapter 4).

To produce electrophoresis, a constant current is passed through the cell. Obviously, it would not be practical to directly compute the electrical field E from the dimensions of the apparatus and the potential difference between the electrodes. Therefore, we make use of Ohm's law. Consider an element of volume, anywhere in the cell, of cross section A and thickness dx. If the potential difference across this slab is dV, then

$$dV = -idR \tag{6.7}$$

where i is the current and dR is the resistance of the slab of solution. (The minus sign corresponds to the convention that the current flows towards the lower potential.) However, we can write dR as dx/KA, where K is the specific conductance of the solution in this particular volume element. Furthermore, $-dV/dx$ is the potential gradient, E. Therefore,

$$E = i/KA \tag{6.8}$$

Thus a knowledge of the current and the cross-sectional area of the cell, together with a separate measurement of the specific conductance, allows the calculation of E in either limb. The current is easily measured. It must be the same throughout the cell, and through the external circuit as well, or charge would accumulate at some point. Since observation of the motion of the boundary can be made with high precision, accurate mobility determinations can be made.

Turning again to Figure 6.4, one can see that there will be two boundaries producing in this arrangement. When the current is turned on, one of these will move upward, and the other will descend. These *ascending* and *descending* boundaries will be nearly, but not exactly, the same in their velocity. This occurs even though the macroion solution was originally dialyzed against buffer so as to make small-ion concentrations as much alike as possible across the boundaries. If a moving boundary is to be observed, the macromolecule must be charged, and the Donnan equilibrium (Chapter 2) will then assure some small difference in diffusible ion concentration. The descending boundary is moving into the original solution, and the specific conductance of that solution can be used in calculating the mobility. On the other hand, as the other boundary ascends, a new solution is created, with macroions added to what was the outside solution in the dialysis. Since the conductivity of these solutions will differ, the velocities will not be the same. That of the descending boundary is almost always used for calculation, since the conductivity of the original protein solution is most easily measured. Actually, the situation is even more complex, for differences in buffer concentration across the protein boundaries lead to differences in mobility within the boundary region, so that in most cases the descending boundary is broader and the ascending boundary sharper than would be expected from simple diffusion broadening. Furthermore, since the areas observed in schlieren photographs correspond to the total concentration changes across the boundaries, these areas will be somewhat different on the two sides. Finally, the concentration inequalities referred to earlier will lead to the presence of small stationary boundaries at the original positions. All of these effects are evident in the example shown in Figure 6.5.

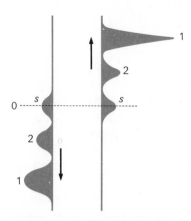

Figure 6.5

Ascending and descending patterns for a hypothetical mixture of two components. Note that peak shapes and mobilities differ in the two limbs and that on each side there is a stationary buffer boundary. See the text.

If there is more than one macromolecular species present in a solution subjected to moving boundary electrophoresis, separate boundaries may be resolved; these may move in the same or opposite directions, depending on the charges. Since the electrophoretic mobility may be markedly changed by as little a difference as one charged group per protein molecule, electrophoresis can provide a sensitive test for protein homogeneity. Even if separate boundaries cannot be resolved, very small amounts of heterogeneity can be detected by the *reversible boundary spreading* test: A boundary is allowed to migrate for some distance, and the current is then reversed; any subsequent sharpening of the boundary cannot be due to diffusion and must reflect heterogeneity.

Moving boundary electrophoresis is the electrophoretic analog of the conventional sedimentation velocity experiment. While it is capable of precision in the determination of mobilities, that precision is often wasted because of the problems in molecular interpretation of mobility values. As a preparative procedure, the moving boundary method leaves much to be desired. Consider the example shown in Figure 6.5. One could recover from this experiment a small amount of pure component 1 from the ascending limb and a small amount of pure component 2 from the descending limb, but most of the mixture would remain unresolved. For preparative purposes, what is clearly required are analogs of the sucrose gradient sedimentation velocity experiment, or the band sedimentation method. Such techniques have been devised and have become exceedingly important, both for preparative and analytical purposes. In fact, their success has so overshadowed that of the moving boundary method that the latter has been relegated, at least for the present, to the status of an obsolete technique.

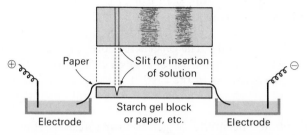

⊕	Paper	Slit for insertion of solution	⊖

Electrode   Starch gel block or paper, etc.   Electrode

Figure 6.6   A schematic view of a starch gel electrophoresis apparatus. The solution is inserted in a slit cut in the gel. If all components are expected to be of the same sign of charge, as in this example, the slit can be placed near one end. The bands are usually revealed by staining. If desired, portions of the block may be cut out and eluted.

Most of these new techniques utilize some semisolid or gelatinous medium (starch gel, as an example) to support the solution being investigated. An example of such a technique is diagrammed in Figure 6.6. These methods possess exactly the same advantages over moving boundary electrophoresis as does density gradient centrifugation over the conventional sedimentation velocity experiment. Since convection is eliminated by the supporting medium, it is possible to start with a thin layer of the solution being tested. Only minute quantities are required, and complete separation of components is possible.

TABLE 6.1   SOME MEDIA FOR ZONAL ELECTROPHORESIS

Medium	Conditions	Advantages	Disadvantages
Paper	Filter paper is moistened with buffer; held between electrodes, as in Figure 6.6	Fast, simple; only small amounts of solute needed	Evaporation and endosmosis may be troublesome
Starch gel	A gel is cast from partially hydrolyzed starch; see Figure 6.6	High resolving power, versatility	Some hindrance to motion for large molecules, some absorption; removal of sample difficult
Agar gel	Like starch gel but more dilute in gel-former	Open structure gives conditions close to free electrophoresis	Negative charge of gel gives high interaction with positively charged solute; endosmosis in alkaline solution
Polyacrylamide gel	Acrylamide is polymerized and cross-linked in small tubes	Excellent resolution, easy to use; endosmosis negligible	Usually restricted to small scale, for resistance is low; currents become large

[a] For more details, see J. L. Bailey (1967).

An almost infinite variety of media and methods have been tried; some of these are summarized in Table 6.1. It is necessary, of course, to have some method of assay to detect the migrating materials, and the conditions are obviously not so satisfactory for precise mobility calculations. The viscosity of the medium is ill defined, interaction of solute with the medium may be of importance, and flow of the solution (endosmosis) through the medium may occur under the influence of the field.

However, these complications should not detract from the enormous importance of these methods. The gel electrophoresis pattern of a protein mixture is reproducible, and the presence of a single band is one of the most sensitive tests for homogeneity. As in the case of gradient centrifugation, the fact that a variety of assay methods can be used is of great advantage in the study of complex mixtures.

An example of the utility of gel electrophoresis is shown in Figure 6.7. The pattern shows the resolution of *isozymes* of enolase (see Wold, this series). Such multiple forms of an enzyme could not be separated at all by centrifugal methods; they differ only in the fact that they are each mixed tetramers of two slightly different kind of subunits.

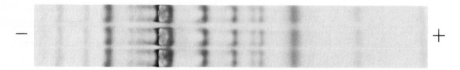

Figure 6.7   Zonal electrophoresis patterns showing the separation of isozymes of Coho salmon enolase. Courtesy of F. Wold.

6.2 ISOELECTRIC FOCUSING

Gel electrophoresis is analogous to the techniques of band sedimentation. There also exists a technique that is the analog of the density gradient sedimentation equilibrium experiment. In this technique, called *isoelectric focusing*, components of different isoelectric points are brought to form stationary bands in a pH gradient. In the sedimentation equilibrium experiment, a particle moves to the point at which the term $(1 - \bar{v}\rho) = 0$ and then stops; in a pH gradient a charged particle will be moved by an electric field until it reaches a point where it is isoelectric. Its net charge will then be zero, and it will stop.

To carry out this kind of experiment, two criteria must be met:

1. Since bands are to be formed, a density gradient must be present in the apparatus to prevent convection. A sucrose gradient is usually employed. Since sucrose is uncharged, it is unperturbed by the electric field.

2. A reasonably stable pH gradient must be set up. This is more difficult. An artificial gradient could be formed (for example, by layering a low-pH solution above the high-pH solution and then lightly stirring at the boundary). But such a gradient would be influenced by the electrodes at its ends, since the ions would be transported quite rapidly when the field was applied. In most cases, such a gradient would change so rapidly that a macromolecular solute would not be able to keep up.

A very ingenious method has been devised to yield stable pH gradients. It is clear that after some electrolysis of a salt solution, the anode and cathode compartments of any electrophoretic apparatus will be acidic and basic, respectively. If a mixture of low-molecular-weight polyampholytes is placed in the intervening region, each will move to its isoelectric region and remain there, stabilizing the pH at a particular value at that point. The most acidic polyampholytes will be concentrated near the anode, and the most basic near the cathode. A stable pH gradient will have been established. Now, any macromolecule in the mixture will migrate to the region at which it is isoelectric. An equilibrium between electrophoresis and diffusion (resembling the equilibrium between sedimentation and diffusion in the sedimentation experiment) will result in concentration of each macromolecular species into a sharp band. Of course, this is not a true equilibrium, for the sucrose gradient must ultimately vanish, and electrolysis is continually going on, but the bands can be stable for rather long times.

The simple apparatus needed for this kind of experiment is shown schematically in Figure 6.8, and a representative separation is given in Figure 6.9.

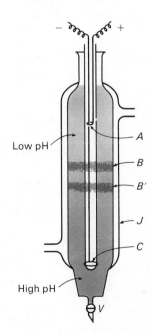

Figure 6.8
A schematic drawing depicting the isoelectric focusing technique. The sucrose gradient is indicated by shading. The anode and cathode are platinum rings, denoted by A and C, respectively. Protein bands are shown formed at B and B'. The column may be emptied through the valve, V. It is cooled by the water jacket, J. After H. Svensson, *Arch. Biochem. Biophys. Suppl.* **1**, 132 (1962).

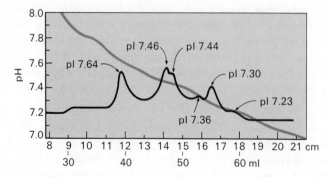

Figure 6.9   Separation of hemoglobin components by isoelectric focusing. The curve with peaks shows the absorption at 250 nm as a function of elutant volume and column position. The monotonic curve is the pH gradient. From A. Haglund, *Sci. Tools*, **14**, 17 (1967).

The technique is remarkably sensitive to very slight charge differences and can be utilized for either analytical or preparative purposes.

6.3   ORIENTATION OF MOLECULES IN ELECTRIC FIELDS

So far, we have discussed the transport of molecules in electric fields. We now turn to another, and equally important effect—the orientation that is produced when the molecules have asymmetric charge distributions. A molecule in which charges $+q$ and $-q$ are separated by a distance d is shown in Figure 6.10. These

Figure 6.10

A dipole in an electric field. While there is no transport, since the net charge is zero, there is a torque tending to orient the molecule in the direction of the field.

charges might represent either ionized groups or simply the asymmetric charge distribution in a polar bond; in either case the molecule would be said to possess a *permanent dipole moment*. Alternatively, the displacement of charge might have been produced by an external electric field; in this case the dipole moment is *induced*. In either event, the dipole moment is a vector whose direction is given by the line connecting the centers of charge and whose magnitude† is defined as

$$\mu = qd \tag{6.9}$$

† Dipole moments are frequently expressed in Debye units. A pair of charges, each of magnitude 1×10^{-10} esu, separated by 1 Å (10^{-8} cm) corresponds to a dipole moment of 1 D. Small polar molecules typically have dipole moments of the order of 1 D; asymmetric macromolecules may have moments of hundreds or thousands of Debye units. Thus orientation experiments become feasible at low fields.

The vector nature of the dipole moment means that we can represent the dipolar character of even a very complex molecule as the single vector sum of all the moments of individual dipoles within the molecule. For example, a synthetic polypeptide in the α-helical conformation (see Barker, in this series) may have a very large permanent moment, because all of the small dipole moments of the amino acid residues along the chain add together, whereas the antiparallel double helix of DNA has no permanent moment but may have a very large induced moment.

A molecule such as is depicted in Figure 6.10 will not be moved by an electric field, since it has no net charge, but it will tend to be oriented. The analysis of this effect is not difficult. Consider the potential energy of the molecule in a field of strength E. The molecule in Figure 6.10 is depicted with its dipole moment at an angle θ with respect to the direction of the field; this places one charge center at a position x_1 and the other at a position x_2 in the field. The potential energy of a charge varies linearly with position in a homogeneous field, and the potential energy of the pair of charges can be considered the sum of the individual energies. Then we may write

$$V = V(x_1) + V(x_2) = -(qx_2 - qx_1)E$$

$$= -q(d \cos \theta)E$$

$$= -\mu E \cos \theta \tag{6.10}$$

Thus the potential energy will be at a minimum when $\theta = 0$, that is, when the molecule is aligned with the field. The magnitude of this potential energy difference is directly proportional to the dipole moment, μ, and to the field strength, E.

We should not expect, of course, that the application of an electric field will produce total alignment of the dipolar molecules. The molecules will be constantly subjected to random interaction with solvent molecules, which will tend to produce random orientation. At equilibrium, a distribution of orientations should result; it is easy to calculate this from the familiar Boltzmann formula. If we let $f(\theta)\, d\theta$ represent the fraction of molecules whose dipole moments make an angle between θ and $(\theta + d\theta)$ with the field direction (see Figure 6.10), we write immediately

$$f(\theta)\, d\theta = Ke^{\mu E \cos \theta / kT} \sin \theta\, d\theta \tag{6.11}$$

Here the quantity in the exponent is just minus the potential energy divided by kT, as always in a Boltzmann distribution. The distribution is weighted by the factor $\sin \theta\, d\theta$ because we are concerned only with the angle of the moment to the field direction and not with rotation *about* that axis. There will be more directions for a molecule to point in a given increment $d\theta$ as θ approaches 90 deg (see Figure 6.11). The constant K could be evaluated by noting that the integral of $f(\theta)\, d\theta$ from zero to π must equal unity. Note that the equation has the same form as Equation (1.10), with $\sin \theta \, d\theta$ playing the role of the degeneracy g.

Having a simple description of the equilibrium orientation of molecules in an electric field allows predictions as to the effect of a field on observable properties. As a first example, we may consider the dielectric constant of the solution. This will depend on the polarizability of the solution, which will in turn depend on the average dipole moment *in the direction of the field*. The Debye equation relates dielectric constant (D) and polarizability (α) in the following way:

$$\frac{D-1}{D+2} = \frac{4}{3}\pi n \alpha \tag{6.12}$$

where n is the number of molecules per cubic centimeter, and

$$\alpha = \alpha_i + \alpha_p \tag{6.13}$$

The terms α_i and α_p represent the *induced* polarizability and the *orientation* polarizability, respectively. They may be distinguished experimentally by the fact that at very high frequencies of alternating electrical fields the molecules are unable to reorient rapidly enough to keep up with the field, and only the induced component is left. For slowly varying fields (up to about 10^5 to 10^6 cps for a typical protein) we may assume that the molecules can follow the field, and use the equilibrium distribution to calculate the average moment ($\bar{\mu}$) in the field direction contributed by the partially oriented molecules. The orientation contribution to the polarizability will then be $\bar{\mu}/E$. If a given molecule makes an angle θ with the field, its contribution is obviously $\mu \cos \theta$; therefore,

$$\bar{\mu} = \frac{\displaystyle\int_0^\pi \mu \cos \theta e^{\mu E \cos \theta / kT} \sin \theta \, d\theta}{\displaystyle\int_0^\pi e^{\mu E \cos \theta / kT} \sin \theta \, d\theta} \tag{6.14}$$

If $\mu E/kT$ is small, a condition almost always satisfied in dielectric constant measurements, the equation reduces to

$$\bar{\mu} \simeq \frac{E\mu^2}{3kT} \tag{6.15}$$

which, when combined with (6.13) yields

$$\alpha = \alpha_i + \mu^2/3kT$$

$$(D-1)/(D+2) = \tfrac{4}{3}\pi n[\alpha_i + \mu^2/3kT] \tag{6.16}$$

This equation enables one to use dielectric constant measurements to determine the permanent and induced dipole moments of macromolecules. Considerable difficulties arise, however, when one attempts to study molecules such as proteins in their natural environment, aqueous solutions. In the first place, measurements are difficult because such solutions conduct electricity, and heating will occur at high fields. Furthermore, in a highly polar solvent such as water, the polarization of the medium modifies the local field, making the validity of the Debye equation dubious. While other equations have been proposed [see C. Tanford (1961), Chapter 2], the results are often compromised by the question of which equation to use. Finally, biological macromolecules often contain weakly ionizing groups, yielding the possibility of a kind of polarization by the migration of ionic charge from one site to another upon the molecule. Since this may be a fairly "slow" process as compared to induced polarization, it is apt to be confused with orientation polarization. For this reason, many of the large permanent dipole moments reported for proteins (see Table 6.2) are suspect.

TABLE 6.2   DIPOLE MOMENTS OF SOME MACROMOLECULES

Substance	Solvent	M	$\mu\,(D)$
Myoglobin	Water	17,000	170
Egg albumin	Water	44,000	250
Horse carboxyhemoglobin	Water	67,000	480
Poly-γ-benzyl-L-glutamate	Ethylene dichloride	70,000	1,040
		102,000	1,570
		153,000	2,460

Partial orientation of the macromolecules in a solution may effect its optical properties. For example, the absorption of polarized light may differ if the direction of polarization is perpendicular or parallel to an electric field. The magnitude of this effect (*electric dichroism*) is easily calculated from Equation (6.11) in the simple case where the transition moment of the chromophore (see Chapter 8) is parallel to the permanent dipole moment of the macro-molecule. As shown in Chapter 8, the extinction coefficient for a chromophore making an angle θ with respect to the direction of polarization of the light is $\varepsilon = \varepsilon_\parallel \cos^2 \theta$, where $\varepsilon_\parallel$ is the extinction coefficient for parallel orientation. Then we may write, for light polarized parallel to the field:

$$\bar{\varepsilon} = \varepsilon_\parallel \frac{\displaystyle\int_0^\pi \cos^2 \theta e^{\mu E \cos \theta / kT} \sin \theta\, d\theta}{\displaystyle\int_0^\pi e^{\mu E \cos \theta / kT} \sin \theta\, d\theta} \simeq \frac{\varepsilon_\parallel}{3}\left[1 + \frac{2}{15}\left(\frac{E\mu}{kT}\right)^2 \right] \tag{6.17}$$

Thus in this case applying an electrical field will increase the absorption of the light; were the transition moment perpendicular to the dipole moment, a diminution in absorption would be observed when the field was turned on.

A closely related phenomenon is *electric birefringence,* or, as it is sometimes called, the *Kerr effect.* Because of the close relationship between absorption and refraction, it is not surprising that oriented solutions of macromolecules will be birefringent, that is, display different refractive indices for light polarized parallel and perpendicular to the electric field. While we shall not go into the theory in detail here, it may be noted that the equations are, as would be expected, very similar to those for electric dichroism. For a rodlike molecule with dipole moment μ, the refractive index difference Δn is given by

$$\frac{\Delta n}{C} = \frac{2\pi E^2}{15 nm}\left(\frac{\mu}{kT}\right)^2 (\alpha_{\parallel} - \alpha_{\perp}) \tag{6.18}$$

Here $\alpha_{\parallel}$ and $\alpha_{\perp}$ are components of the optical polarizability parallel and perpendicular to the molecular axis, m is the molecular mass, n is the refractive index, and C is the concentration. Both electric dichroism and electric birefringence are exceedingly powerful techniques for studies of the anisotropy of rodlike macromolecules. The former technique is possibly the more potentially fruitful, for it allows in principle the investigation of the relative orientation of various chromophoric groups on a given molecule, if these absorb at different wavelengths.

Further insight into the properties of macromolecules can be obtained by studying the *relaxation* of a partially oriented system to a state of random orientation when the field is suddenly turned off. Here we are concerned with a new type of nonequilibrium process, *rotational diffusion.* If we have a large collection of anisotropic molecules (let us assume them to be rods, for convenience), we may describe their distribution of orientations by a diagram such as Figure 6.11. We are concerned only with the orientation with respect

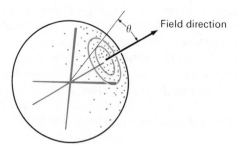

Field direction

Figure 6.11

Statistics of molecular orientation in an electrical field. Each point on the sphere surface represents the end of a rodlike molecule at a given instant. For random orientation the points would be uniformly distributed over the sphere surface. This is a very nonrandom distribution. The function $f(\theta)\, d\theta$ represents the fraction of molecules whose end points lie in the annular ring between θ and $\theta + d\theta$.

to a particular axis, z. Imagining all of the rods to be of the same length, and with their centers at the same point (0), we see that their ends will form a pattern of points on the surface of a sphere. The distribution $f(\theta, t)\, d\theta$ will then represent the fraction of points that lie in the annular ring between θ and $\theta + d\theta$, at a particular time t. The random rotational diffusion of orientations will then be represented by the random diffusion of points on the surface of a sphere. It is not hard to see that the process will in many ways resemble the process of translational diffusion. There will be an analog of Fick's first law for the flow of points across the circle defined by a particular θ:

$$J(\theta, t) = -\Theta\left(\frac{\partial f(\theta, t)}{\partial \theta}\right)_t \tag{6.19}$$

The analog of Fick's second law is

$$\frac{\partial f(\theta, t)}{\partial t} = \Theta \frac{\partial^2 f(\theta, t)}{\partial \theta^2} \tag{6.20}$$

In these equations, the quantity Θ is a *rotational diffusion coefficient*; like the translational diffusion coefficient it will depend on the size and shape of the macromolecule.† For example,

$$\Theta = RT/8\,\mathcal{N}\pi\eta R^3 \qquad \text{(sphere of radius } R) \tag{6.21}$$

$$= \frac{3RT}{16\,\mathcal{N}\pi\eta a^3}\left(2\ln\frac{2a}{b} - 1\right) \qquad \text{(prolate ellipsoid, length } 2a, \text{ thickness } 2b) \tag{6.22}$$

The process of relaxation from partial orientation can now be described in the following way: Initially, after "long" (several milliseconds) application of the field, the distribution of orientations will look somewhat like that shown in Figure 6.11; it will be described by Equation (6.11). When the field is suddenly removed, the rotatory diffusion process will relax the distribution to a random one, in which the end points in Figure 6.11 become uniformly distributed over the sphere. The process is very much like what would occur if we could suddenly turn off the centrifugal field on a system at sedimentation equilibrium. Without working out the result, we can guess that a property such as the birefringence or dichroism should relax exponentially with time toward zero. In fact, the equation for a property X that depends on the orientation is

$$X(t') = X_{eq}e^{-6\Theta t'} \tag{6.23}$$

where t' is the time after the field is turned off and X_{eq} the value at equilibrium orientation and subsequent relaxation. Thus, in addition to the other information provided, electric birefringence or dichroism studies should yield rotational diffusion coefficients and hence data on molecular dimensions. An apparatus for electric dichroism measurements is diagrammed in Figure 6.13.

It must be understood that the discussion presented here applies to the simple case of a cylindrically symmetrical, rodlike macromolecule. Most real

† Note that the rotational diffusion coefficient depends on the *cube* rather than the first power of a dimension of the macromolecule. This makes it exceedingly sensitive to the lengths of asymmetric molecules.

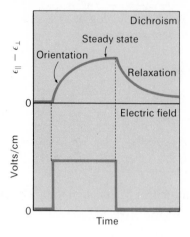

Figure 6.12

Orientation and relaxation in an electric dichroism experiment.

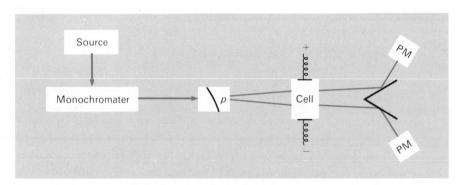

Figure 6.13   A schematic drawing of the apparatus used to obtain the kind of data shown in Figure 6.12. Monochromatic light is passed through a Wollastron prism (*p*), which splits the beam into two divergent beams, one polarized parallel and the other perpendicular to the field. These beams are directed to two photomultipliers, and the intensity difference is measured.

molecules will not possess such symmetry, and in the general case, *three* rotational diffusion coefficients, to describe rotation about the three axes, become necessary. Needless to say, all of the theories become correspondingly more complex.

PROBLEMS

1. Data from an electrophoresis experiment with γ-globulin: The descending boundary moved 1.40 cm in 135.6 min; specific conductivity of protein = 2.32×10^{-3} ohm^{-1}/cm; cross-sectional area of cell = 0.75 cm^2; current = 20 mA.

(a) Calculate the electrophoretic mobility in square centimeters per second per volt.

(b) If the specific conductivity of the solution is found to be 2.15×10^{-3} ohm^{-1}/cm, how far will the ascending boundary have moved in 135.6 min?

*2. Consider a long rigid structure, such as DNA, that has negatively charged groups spaced regularly along the rod. Using approximations suggested by Figure 4.2, write an equation that would describe the dependence of electrophoretic mobility on molecular weight. (*Note:* Only the functional dependence, not the values of parameters, is required.)

3. At pH 7.0 in 0.1 ionic strength buffer at 4°C, normal hemoglobin has a mobility of about -0.2×10^{-5} cm²/sec-V, whereas sickle-cell hemoglobin has a value of about $+0.3 \times 10^{-5}$ cm²/sec-V. Deduce the charge on each type of molecule, using (a) The approximate Equation (6.4), and (b) The more exact expression (6.5). You may assume the hemoglobin molecule, in each case, to be a sphere of radius 32 Å, the dielectric constant to be 80, and the ionic strength to be 0.1.

*4. A combination ultracentrifuge-electrophoresis apparatus might be made by placing electrodes at the bottom and meniscus of a solution in the ultracentrifuge cell. Derive an expression for the velocity of motion of a charged molecule under the combined influences of these fields. For electrodes spaced at 7.00 and 6.00 cm from the center of rotation of a rotor spinning at 60,000 rpm, what voltage difference would be required to hold a bovine serum albumin boundary at 6.50 cm from the center of rotation? The sedimentation coefficient of bovine serum albumin can be taken as 4.5×10^{-13} sec for this calculation, and the charge on the molecule is $+10$ (proton charge units).

*5. From the data given in Table 6.2, develop evidence that poly-γ-benzyl-L-glutamate is rodlike in ethylene dichloride.

6. Derive the approximate result given by (6.17). You will find that this approximation is valid only at low field strengths.

7. Calculate the percent change in extinction coefficient according to (6.17) for a field 10,000 V/cm, assuming a dipole moment of (a) 10 D, and (b) 1000 D.

8. For the conditions listed in Problem 7, calculate the fraction of molecules oriented between 0 and 90 deg to the field direction.

9. Tobacco mosaic virus has a rotational diffusion coefficient of about 333 sec⁻¹.
(a) Calculate the time required for a quantity such as electric birefringence to decay to $1/e$ of its initial value. This is known as the *relaxation time.*
(b) Estimate the length of the TMV particle from the value of Θ, assuming its diameter to be 180 Å and the temperature to be 20°C.
(c) Estimate the relaxation time for an end-to-end dimer of TMV. Such particles are sometimes observed.

REFERENCES

Electrophoresis (Moving Boundary)

Alberty, R. A.: in *The Proteins*, H. Neurath and K. Bailey (eds.), 1st ed., Vol. I, Part A, Academic Press, Inc., New York, 1954.

Overbeek, J. Th. G.: *Advan. Colloid. Sci.*, **3**, 97 (1950).

Electrophoresis (Zonal)

Bailey, J. L.: *Techniques in Protein Chemistry*, 2nd ed., American Elsevier Publishing Company, Inc., New York, 1967, Chap. 10.

Dipole Moments

Oncley, J. L.: in *Proteins, Amino Acids, and Peptides*, E. J. Cohn and J. T. Edsall (eds.), Reinhold Publishing Corporation, New York, 1943, Chap. 22.

SEVEN | VISCOSITY

The measurement of diffusion, sedimentation, and electrophoretic mobility provide considerable information about the size, shape, and gross conformation of dissolved macromolecules. However, it should be obvious that this information is far from complete and, in some cases, is beset with ambiguities in interpretation. To supplement these kinds of study, viscosity measurements are often used. To state their utility in an oversimplified way: The contribution of a macromolecular solute to the viscosity of a solution depends primarily on the effective volume occupied by the macromolecules. This is to be contrasted to the frictional coefficient, which depends on linear dimensions (a radius, for example; see Stokes' law). In this chapter we shall examine briefly the mechanics of viscous flow and show what studies of flowing solutions can tell about the macromolecular dimensions.

7.1 VISCOUS FLOW

The resistance of fluids to flow is measured by their *viscosity*. To provide a precise definition of this quantity, it is necessary first to examine the mechanics of viscous flow. Consider the simple arrangement shown in Figure 7.1. A liquid is held between two infinite (very large) parallel plates. One of these is at rest; the other is pushed in the x direction with a constant velocity v. It is assumed that the infinitesimal layer of liquid in contact with each plate "sticks" to it, so that the intervening layers of fluid must slide over one another

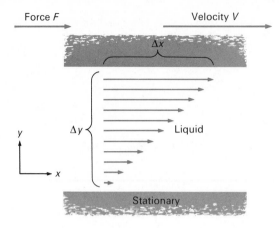

Figure 7.1

A schematic representation of the shearing of a Newtonian liquid between parallel plates. The arrows represent fluid velocities at different values of the coordinate y.

in the manner shown. It is the degree of resistance to such sliding that is measured by the viscosity.

The kind of deformation that is experienced by the liquid sample in Figure 7.1 is referred to as shear. The *shear strain* at any moment is defined in this case as $\Delta x/\Delta y$. (Obviously, in this simple example the shear strain is everywhere uniform; were it not, we could define a shear strain at any point as dx/dy.) The *shear stress* is defined as the force pushing the top plate in the x direction, divided by the area (A) of the plate in contact with the sample. A liquid, when subjected to a constant shear stress, will shear at a constant *rate* as long as the stress is maintained.† For most liquids (Newtonian liquids) the relation between shear stress σ and rate of shear strain is very simple:

$$\sigma = \eta \frac{d}{dt}\left(\frac{dx}{dy}\right) \tag{7.1}$$

In this equation, η is the viscosity of the liquid; this expression may be taken as a definition of the viscosity. It may be written in a different way by changing the order of differentiation in (7.1):

$$\sigma = \eta \frac{d}{dy}\left(\frac{dx}{dt}\right) = \eta\beta \tag{7.2}$$

Since dx/dt is the velocity of the liquid layer at height y, this formulation states that the viscosity is the ratio of the shearing stress to the gradient in fluid velocity (β). These definitions specify the dimensions of η. Since stress is in

---

† The response to shear stress is very different for solids and liquids and provides a mechanical distinction between these states. An ideal solid, subjected to a constant shear stress, will simply assume an equilibrium value of the shear strain. Upon removal of the stress, the strain will vanish. Some substances exhibit combinations of the properties of both ideal solids and liquids. These are called *viscoelastic* materials; examples include most solid plastics, concentrated polymer solutions, and the like.

dynes per square centimeter, viscosity has the dimensions of grams per centimeter-second; these units are called poise. The viscosity of water at room temperature is about 10^{-2} P, or 1 CP.

The general conditions of flow described above correspond to *laminar* flow. They will usually be met in such experimental conditions as flow between concentric rotating cylinders or through a capillary tube so long as the rate of shear is not too great. At very high shear rates, turbulence will commence; turbulent flow is much more complex and will not be discussed here. The Newtonian viscosity law [Equation (7.1) or (7.2)] describes the flow of almost all low-molecular weight liquids and some macromolecular solutions as well. The complications encountered when the shear rate is a more complex function of stress than given by (7.1) will be discussed later.

7.2   THE VISCOSITIES OF MACROMOLECULAR SOLUTIONS

The interest of the biophysical chemist in viscosity measurements stems from the fact that the addition of some macromolecular solutes may greatly alter the viscosities of their solutions. While many theories of the viscosity of macromolecular solutions have been developed, all have their origin in the pioneering work of Albert Einstein. In 1906, Einstein developed the equation for the viscosity of a solution containing rigid, spherical solute particles. While the details of the derivation involve much sophisticated hydrodynamics, the general tenor of the argument can be presented easily. We begin by pointing out still another possible definition of viscosity: Viscosity is a measure of the rate of energy dissipation in flow. From Figure 7.1 it is clear that we may write the shear rate as $v/\Delta y$; if we multiply both sides of Equation (7.2) by this, we obtain

$$\frac{F}{A\,\Delta y}\frac{dx}{dt} = \eta\left(\frac{v}{\Delta y}\right)^2 \tag{7.3}$$

where the shear stress has been written out as a ratio of force to area. But $F\,dx$ is the energy expended in time dt in deforming the sample of fluid, and $A\,\Delta y$ is the volume of the fluid. Thus the left side of Equation (7.3) represents the rate of dissipation of energy per unit volume of fluid; this is seen to be proportional to the square of the shear rate, the factor of proportionality being η. That is

$$\frac{dE}{dt} = \eta\left(\frac{v}{\Delta y}\right)^2 \tag{7.4}$$

Now consider a viscometer, like that in Figure 7.1, wherein we first measure the viscosity of a solvent by determining the shearing force necessary to produce a given rate of shear. We then introduce into the liquid a number of rigid particles; let us say that all together they occupy a fraction ϕ of the solution volume. We shall now find that a greater force is required to maintain the same shear rate:

$$\left(\frac{dE}{dt}\right)_s > \left(\frac{dE}{dt}\right)_0 \tag{7.5}$$

The rate of energy dissipation per unit volume is greater because in the solution there is a smaller volume of fluid in which the same overall deformation must be accomplished—the solute particles themselves cannot be deformed. This leads us to guess that the new rate of energy dissipation must be related to the rate in pure solvent by an equation of the form

$$\left(\frac{dE}{dt}\right)_s = \left(\frac{dE}{dt}\right)_0 (1 - v\phi)^{-1} \simeq \left(\frac{dE}{dt}\right)_0 (1 + v\phi) \tag{7.6}$$

which yields for the ratio of the viscosities

$$\frac{\eta_s}{\eta_0} = 1 + v\phi \tag{7.7}$$

Here v is a numerical factor, whose value we do not know. By a (vastly) more rigorous treatment of the problem, Einstein showed that for spherical particles $v = \frac{5}{2}$. For particles of other shapes, in dilute solutions, one finds an equation of the form of (7.7) with different values for v. Figure 7.2 shows the factor v as a function of axial ratio for prolate and oblate ellipsoids of revolution.

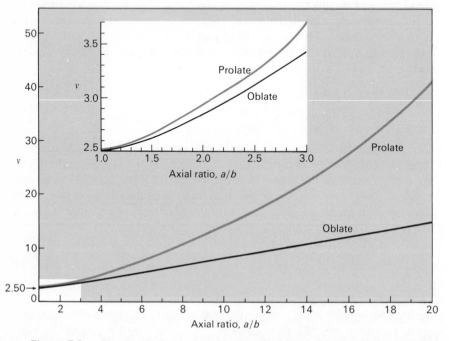

Figure 7.2   The viscosity factor v as a function of axial ratio for ellipsoids for revolution.

Equation (7.7) applies only in dilute solutions, for even its rigorous derivation neglects the possible effects of solute–solute interaction. These effects will give rise to quadratic and higher terms in the concentration, so that, in general, one should write

$$\eta_r = \frac{\eta_s}{\eta_0} = 1 + v\phi + \kappa\phi^2 + \cdots \tag{7.8}$$

Here we have introduced the term *relative viscosity* (η_r) for the ratio of solution to solvent viscosity. If we subtract unity from this, we obtain the *specific viscosity*, which is a measure of the fractional change in viscosity produced by adding the solute:

$$\eta_{sp} = \eta_r - 1 = \frac{\eta_s - \eta_0}{\eta_0} = v\phi + \kappa\phi^2 + \cdots \tag{7.9}$$

Generally, it is awkward to work in a volume fraction concentration scale, for one usually prepares solutions by weighing solute into a given volume of solution. We can rewrite (7.9) by noting that $\phi = Cv$, where C is in grams per milliliter, and v is the specific volume of the solute. Then

$$\eta_{sp} = vvC + \kappa v^2 C^2 + \cdots \tag{7.10}$$

or

$$\frac{\eta_{sp}}{C} = vv + \kappa v^2 C + \cdots \tag{7.11}$$

The limit of η_{sp}/C as $C \to 0$ is termed the *intrinsic viscosity*,† $[\eta]$; it will depend on the properties of the isolated macromolecules, since effects of interaction have been eliminated by extrapolation. According to Equations (7.9) and (7.11) one may determine $[\eta]$ by measuring the relative viscosity at a number of solute concentrations and extrapolating η_{sp}/C to infinite dilution (see Figure 7.3). From Equation (7.11) we see that

$$[\eta] = vv \tag{7.12}$$

which says that the intrinsic viscosity of a macromolecule depends on its shape (through v) and its specific volume. The specific volume, however, is not that of the anhydrous polymeric material but must include, to some extent at least, solvent that is attached to the molecule. The value of 2.5 that v takes for spherical particles is a minimum; any deviation from sphericity leads to a larger value (see Figure 7.2). For proteins, the *anhydrous* specific volume is

---

† Note that relative viscosity, specific viscosity, and intrinsic viscosity do not have the dimensions of viscosity at all. The first two are dimensionless ratios, while $[\eta]$ has units of milliliters per gram.

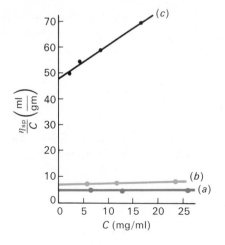

Figure 7.3

Determination of the intrinsic viscosity of crab hemocyanin in various solution environments. (*a*), Data for the native hemocyanin ($M \simeq 1 \times 10^6$) at neutral pH. (*b*), Subunits of the hemocyanin ($M \simeq 7 \times 10^4$) at pH 10.6. (*c*), Unfolded subunits in $6M$ guanidine hydrocloride, pH 10.6.

about 0.75 in most cases, so that we may estimate the minimum value for the intrinsic viscosity (spherical proteins, no hydration) to be about 2 ml/g. As can be seen from the results collected in Table 7.1, a number of proteins approach this value, testifying to their somewhat compact shape. On the other hand, such elongated, rodlike molecules as DNA and myosin exhibit much higher values.

The principal difficulty in interpreting intrinsic viscosity data for globular proteins lies in the fact that both hydration and asymmetry are contributing to the same quantity. If one assumes that water of hydration has about the same density as ordinary water, then a hydration of δ g of water/g of dry protein

TABLE 7.1   VISCOSITY BEHAVIOR OF MACROMOLECULES

	Substance	M	$[\eta]\,(ml/g)$	$\beta \times 10^{-6}$	$\Theta\,(sec^{-1})$
Globular particles	Ribonuclease	13,683	3.4	2.01	
	Serum albumin	67,500	3.7	2.04	6.4×10^5
	30S *E. Coli* Ribosomes	900,000	8.1		
	Bushy stunt virus	10,700,000	3.4		
Rods	Fibrinogen	330,000	27	2.15	4.0×10^4
	Myosin	440,000	217		
	Tobacco mosaic virus	39,000,000	36.7	2.61	3.4×10^2
	Poly-γ-benzyl-L-glutamate[a]	340,000	720		
Coils	Poly-γ-benzyl-L-glutamate[b]	340,000	184		
	Serum albumin[c]	67,500	52		
	Myosin subunits[c]	197,000	93		

[a] Measured under conditions such that the molecule is an α-helix.

[b] Measured under conditions such that the molecule is a random coil.

[c] Measured in $6M$ guanidine hydrochloride; the molecule is a random coil.

will yield $v \simeq \bar{v}(1 + \delta \bar{v}_0/\bar{v})$, where $\bar{v}$ is the partial specific volume of anhydrous macromolecular solute, and $\bar{v}_0$ is the partial specific volume of water. Therefore,

$$[\eta] = v(\bar{v} + \delta \bar{v}_0) \tag{7.13}$$

If δ is known, the axial ratio can be calculated from v (although one still does not know whether the molecule is a prolate or oblate ellipsoid or even if it remotely resembles an ellipsoid). On the other hand, given v from separate knowledge about shape, one may obtain δ. Neither of these procedures has turned out to be especially fruitful, although sometimes the combination of a number of kinds of measurement (intrinsic viscosity, frictional coefficient, and rotatory diffusion coefficient, for example) can provide enough redundancy of information to allow intelligent guesses about both shape and hydration.

A more direct attack on this problem is provided by the formulation of H. A. Scheraga and L. Mandelkern (1953). These authors have combined sedimentation coefficient (s), intrinsic viscosity, and molecular weight to yield the equation

$$\frac{\mathcal{N}s[\eta]^{1/3}\eta_0}{(100)^{1/3}M^{2/3}(1 - \bar{v}\rho)} = \beta(a/b) \tag{7.14}$$

where $(1 - \bar{v}\rho)$ is the familiar buoyancy term in sedimentation theory, $\mathcal{N}$ is Avogadro's number, η_0 is solvent viscosity, and $\beta(a/b)$ is a function of the shape of the particle. A graph of β versus axial ratio (a/b) is shown in Figure 7.4.

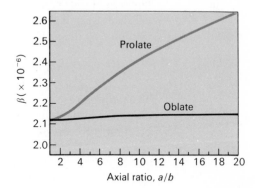

Figure 7.4

The factor β as a function of axial ratio for ellipsoids of revolution. The limiting value of β for an infinitely thin oblate ellipsoid is 2.15×10^6. Thus any molecule with a larger β must be prolate.

It will be noted that this combination possesses the peculiar advantage that β behaves rather differently for prolate and oblate particles. Thus, given supplementary information, one should be able to make a decision about particle shape. That all is not straightforward, however, is indicated by the values of β listed in Table 7.1. A number of apparently reliable results fall below the minimum value of 2.12×10^6 predicted by the theory. The Scheraga–Mandelkern equation has actually been used most often as a way for estimating the molecular weights of substances such as very high-molecular-weight nucleic acids, for which the more conventional techniques are not very satisfactory. The relative insensitivity of β to shape makes a good guess for β possible for this calculation.

The intrinsic viscosities of solutions of randomly coiled macromolecules are often found to be vastly larger than those for compact structures (see Figure 7.3). The reason for this is not hard to see, in a qualitative way. Such molecules occupy a very large effective volume in solution, and intrinsic viscosity is primarily a measure of particle volume. We can arrive rather easily at an expression for $[\eta]$ by assuming that the effective radius of the polymer coil will be of the same order of magnitude as the radius of gyration, R_G. If Equation (7.12) is rewritten on a volume-per-molecule rather than a volume-per-gram basis, it becomes

$$[\eta] = v \mathcal{N} \frac{v_m}{M} \tag{7.15}$$

where v_m is the molecular volume and M the molecular weight. Now if we assume that $v_m = K'' R_G^3$, where K'' is a constant, we obtain an expression of the form

$$[\eta] = K' \frac{R_G^3}{M} \tag{7.16}$$

with K' another constant.

For a series of samples of a given polymer, the radius of gyration will be a function of the chain length and hence of the molecular weight. If the polymer is dissolved in a θ solvent (see Chapter 2), the radius of gyration will be proportional to the square root of the molecular weight (C. Tanford, 1961):

$$[\eta] = K M^{1/2} \tag{7.17}$$

Thus in this case the intrinsic viscosity will depend on the square root of the molecular weight. If the solvent is "better" than a θ solvent, the radius of gyration will increase more rapidly with M. Then an equation of the form

$$[\eta] = K M^a \tag{7.18}$$

will obtain, where a is some number greater than one half. These predictions are supported by the data summarized in Table 7.2. Note that whenever a θ solvent is used, $a = 0.5$, as predicted. Equations of this type are exceedingly useful to the polymer chemist, for they allow an easy determination of molecular weights from intrinsic viscosity measurements. However, the parameters K and a must always be determined empirically for each solvent–solute system, using for calibration samples of known molecular weight. This dependence of the intrinsic viscosity on M, it should be emphasized, is peculiar to random-coil polymers and results from the way in which their effective radius depends on M. For compact particles there is no direct dependence of $[\eta]$ on molecular weight. Equation (7.15) emphasizes that $[\eta]$ depends on the ratio of molecular volume to molecular weight. As shown in Table 7.1, some virus particles have intrinsic viscosities as small as the smallest proteins, while smaller asymmetric molecules may have large values of $[\eta]$.

TABLE 7.2 **INTRINSIC VISCOSITY MOLECULAR-WEIGHT RELATIONSHIPS FOR POLYMERS**: $[\eta] = KM^a$

Polymer	Solvent	$T\,(^\circ C)$	$K \times 10^2$	a
Polystyrene	Benzene	25	0.95	0.74
	Cyclohexane	34^a	8.1	0.50
Polyisobutylene	Benzene	24^a	8.3	0.50
	Cyclohexane	30	2.6	0.70
Amylose	0.33 N KCl	25^a	11.3	0.50
Poly-γ-benzyl-L-glutamate	Dichloro acetic acid	25	0.28	0.87^c
	Dimethyl formamide	25	1.4×10^{-5}	1.75^d
Various proteins	6 M guanidine hydrochloride $+0.1\ M\ \beta$-mercaptoethanol		0.716^b	0.66

[a] θ solvents.

[b] The equation is given in the form $[\eta] = Kn^a$, where n is the number of amino acid residues. See C. Tanford et al. (1967).

[c] Coil form.

[d] Helical form.

When a globular protein molecule is denatured, its conformation is altered to something rather like that of a random coil. There is an accompanying increase in the effective volume in solution, so that $[\eta]$ generally increases. This can be a convenient way to study denaturation processes, as is illustrated in Figure 7.5. C. Tanford et al. (1967) have investigated a number of proteins in guanidine-HCl solutions. These results attest to the unfolded state of the polypeptide chain (see Table 7.2).

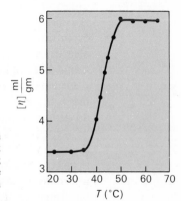

Figure 7.5

The change in intrinsic viscosity at pH 2.8 of the protein ribonuclease upon thermal denaturation. The value at low temperature is typical for a compact globular protein; the increase reflects the partial uncoiling of the molecule at high temperatures. The process is reversible, for upon cooling the value of $[\eta]$ again decreases.

7.3   ORIENTATION IN VISCOUS FLOW

If an elongated macromolecule is placed in a liquid subjected to shear, there will be a torque exerted that tends to orient the molecule so that its long axis is parallel to the direction of flow (see Figure 7.6). In many respects, the situation is similar to that described for a dipole in an electric field (Chapter 6). The random Brownian rotational motion of the molecule will oppose such orientation, and the effect of a constant shear rate will be to produce an equilibrium distribution of orientation states. Note, however, that this is not a true equilibrium but rather a steady state, since energy is constantly being dissipated in the medium.

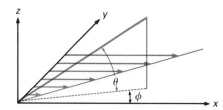

Figure 7.6

The coordinate system used to describe the orientation of an elongated molecule in a shear gradient.

The theory is a good deal more complicated than that for orientation in an electric field. In the first place, as can be seen from Figure 7.6, it is necessary to describe the orientation with respect to two angles, ϕ and θ. Most results that have been obtained are given in terms of series solutions, valid only at low rates of shear. For the fraction of molecules with orientation angles between θ and $\theta + d\theta$ and ϕ and $\phi + d\phi$, the approximate expression has been obtained:

$$\rho(\phi, \theta) = K\left[1 + \frac{\beta}{\Theta} \frac{A \sin 2\phi \sin^2 \theta}{4} + \cdots\right] \tag{7.19}$$

This equation applies specifically to ellipsoids of revolution with semiaxes a and b. The shear gradient [see Equation (7.2)] is denoted by β, Θ is the rotational diffusion constant, and A depends on the axial ratio of the molecule: $A = (a^2 - b^2)/(a^2 + b^2)$. The fact that powers of A are involved in the series expansion means that the result is restricted not only to small shear rates but also to relatively small axial ratios. Calculations have been carried to higher terms, but application of the theory rapidly becomes quite cumbersome.

Orientation by shear manifests itself in optical properties of the medium, much as does orientation by electric fields. Shear orientation possesses the advantage that it is not restricted to macromolecules with large dipole moments or large polarizability but suffers in application from the complexity of the theory and the fact that relaxation experiments, which have proved so fruitful in electric birefringence studies, are much more difficult. However, much useful information has been gained from the shear analogs of electric dichroism and birefringence. Results from shear dichroism studies of F-actin are shown in

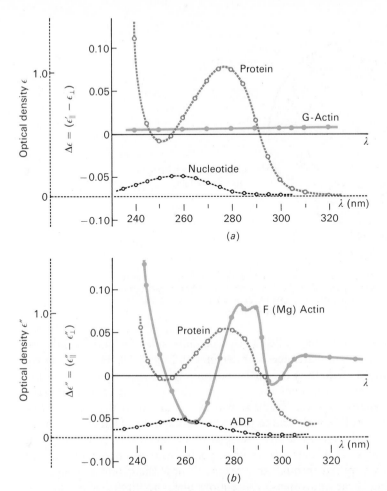

Figure 7.7 Flow dichroism of (*a*) globular actin and (*b*) fibrous actin. In each case the absorption spectrum is shown by the open circles with broken lines, and the dichroic spectrum by the closed circles and full lines. Polymerization of the nondichroic, globular actin monomers leads to fibers that exhibit strong dichroism. The results show that the aromatic residues of the protein are oriented preferentially with transition moments parallel to the polymer axis, while the nucleotide transition moments are perpendicular to the axis. From A. Wada (1964). Reprinted with permission of author and publisher (John Wiley & Sons, Inc. Copyright © 1964.)

Figure 7.7. This fibrous protein exists as a complex with adenosine diphosphate. When the solution is oriented by shear, the directions observed for the polarization of the absorption by the ADP and the aromatic side chains of the protein differ, with the result that the dichroism changes in sign in the vicinity of 275 nm.

Birefringence in shear is usually not studied by direct measurement of the refractive index difference for light polarized in different directions but by

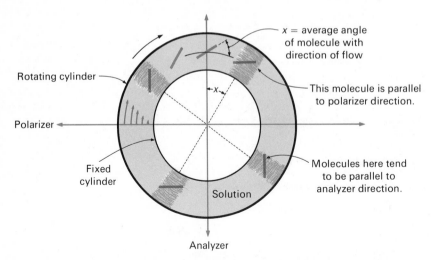

Figure 7.8   This is a schematic view, looking down on an apparatus for flow birefringence experiments. The orientations of "typical" molecules are shown by the rods in the solution. Note that extinction occurs only when the molecular axes are parallel to either the polarizer or analyzer directions.

determination of a quantity called the *extinction angle*. To understand this, examine the experimental arrangement depicted in Figure 7.8. Here are shown concentric cylinders with their axes perpendicular to the paper and the solution placed in the annular gap between them. If one cylinder is rotated and the other is held at rest, a shear gradient will be produced in the liquid. Now suppose that we look down through the space between the cylinders and place a polarizer and an analyzer above and below the apparatus, respectively. If these are rotated so that their directions of polarization are perpendicular, no light will be transmitted through the system. If now the cylinder is made to turn, producing shear and orienting the molecules, the birefringence so produced will allow light to once more be transmitted—but not everywhere. At four positions around the circle the average orientation of the macromolecules will be parallel to the direction of polarization of either the analyzer or the polarizer. At these points the solution acts simply as an extension of the polarizer or analyzer, so light does not get through. Thus we shall observe a crosslike pattern, as shown in Figure 7.8. The angle χ that this pattern makes with the polarizer directions will depend on the orientation of the particles. It is now obvious why this angle is called the extinction angle. If we have a theory for the distribution of molecular orientations, this angle can be predicted. Using the distribution given by (7.19), χ becomes

$$\chi = 45° - \frac{1}{12} \frac{\beta}{\Theta} \frac{180}{\pi} + \cdots \qquad (7.20)$$

Thus observation of the flow birefringence in this way allows the evaluation of the rotational diffusion constant from the limiting slope of χ versus β at low β. As pointed out in Chapter 6, the rotational diffusion constant is a useful quantity for it depends primarily on the length of a rodlike molecule. Some results obtained from orientation in shear are summarized in Table 7.1.

At low rates of shear, χ should approach 45 deg, though its observation becomes difficult, as the pattern fades with the decrease in orientation. The limit of χ at infinite shear, where all of the molecules should be oriented parallel to the direction of flow, should be 0 deg, but it is difficult to reach very high shear rates without turbulence, which destroys the orientation.

A final effect of the orientation of macromolecules in shear is a decrease in the viscosity itself. We have so far assumed that Newton's law [Equation (7.1)] is of general validity, but it is easy to see why solutions of elongated macro-molecules might behave as *non-Newtonian* liquids. As the solute molecules are aligned by the gradient, the resistance that they contribute to the sliding of layers of liquid past one another decreases. Theories have been developed to predict this effect, again in terms of the distribution of orientations. A very clear example of the effect of particle shape on non-Newtonian behavior is shown in Figure 7.9. The polypeptide molecules in the helical conformation orient easily, as one would expect them to. The same sample when placed in a solvent in which the random-coil structure is favored demonstrates a much smaller degree of orientation.

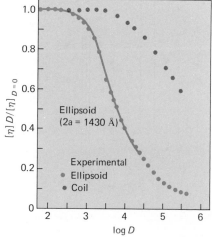

Figure 7.9

The non-Newtonian behavior of poly-γ-benzyl-L-glutamate in *m*-cresol (open circles) and dichloroacetic acid (filled circles). In the former case the polypeptide is a rodlike helix. From J. T. Yang (1958) Reprinted with permission of the author and publisher (Academic Press, copyright © 1958).

The non-Newtonian nature of many polymeric solutions can be an experimental nuisance in the determination of intrinsic viscosity. If one employs a simple capillary viscometer (see Section 7.4) with a moderately high rate of shear for the study of, say, DNA solutions, a value for the intrinsic viscosity may be obtained that is much smaller than the value that would be found at low shear rate. In such studies it has become routine to extrapolate $[\eta]$ to zero rate of shear.†

---

† In the case of very high-molecular-weight DNA, a further complication would appear at high shear. It is found that the viscosity *irreversibly* decreases as successive experiments are performed. This is because high shear gradients can physically break very long molecules. For this reason, high-molecular-weight DNA solutions should not be run through a narrow pipette or syringe.

7.4   MEASUREMENT OF VISCOSITY

As can be seen from Equation (7.9), the determination of the intrinsic viscosity of a macromolecule requires the measurement of the relative viscosities of dilute solutions. Since the relative viscosity of a protein solution containing 1 mg/ml may be as low as 1.003, it is obvious that very high precision in the viscosity measurement is required if the value of $[\eta]$ is to have any reliability at all. Capillary viscometers are well suited to this purpose, for while they do not easily yield absolute viscosity values, they can provide a very accurate comparison of the viscosity of a solution and solvent.

The simplest and most common type of capillary viscometer is the Ostwald viscometer, shown in Figure 7.10(a). One simply measures the time required for a given volume of liquid (as measured by marks on the upper bulb) to flow through the capillary. The volume rate of flow of a liquid through a capillary of radius a and length l, when driven by a pressure P, is given by Poiseuille's

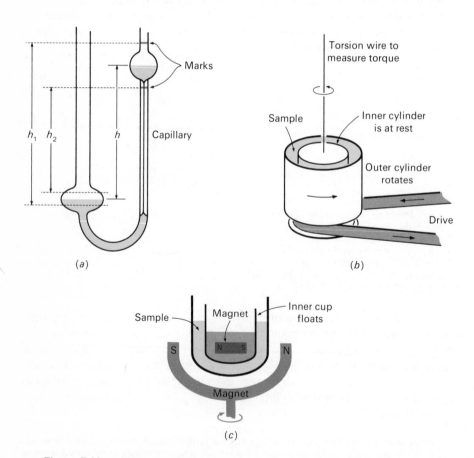

Figure 7.10   Viscometers. See the text.

law:

$$\frac{dV}{dt} = \frac{\pi a^4 P}{8\eta l} \tag{7.21}$$

Here dV/dt is in cubic centimeters per second and η is the viscosity. The liquid is driven, of course, by the hydrostatic pressure head. $P = hg\rho$ (g is the gravitational constant and ρ the liquid density). The height h varies slightly between limits h_1 and h_2 as the bulb empties. Integrating Equation (7.21) yields an expression for the flow time:

$$t = \frac{\eta}{\rho}\left(\frac{8l}{\pi g a^4}\int_{h_1}^{h_2}\frac{dV}{h}\right) \tag{7.22}$$

It is best to regard all of the factors in parentheses on the right of (7.22) (including the integral) as instrument constants. Then, for the measurement of a relative viscosity, we may use the simple expression

$$\frac{\eta_s}{\eta_0} = \frac{t_s}{t_0}\frac{\rho_s}{\rho_0} \tag{7.23}$$

where the subscript zeros refer to solvent properties. According to this equation, the calculation of relative viscosity follows directly from the ratio of flow times if the density ratio is known. The measurements can be made very precisely, but it should be pointed out that good temperature control is needed, for the viscosity is highly temperature-dependent. (A difference of 1°C makes about a 2 percent difference in the viscosity of water; this is a difference comparable to many specific viscosities.)

The capillary viscometer has the disadvantage that the rate of shear cannot be very easily changed or even precisely specified. The pattern of flow through a capillary tube is such that the shear gradient is greatest at the walls and zero in the center. This means that only an average shear gradient can be specified and that it is determined by the apparatus constants and decreases during the flow (as the driving head falls). For measurements of non-Newtonian liquids, these complications are annoying. While capillary viscometers have been built that have several bulbs at different heights to provide several average shear rates, or are driven by an externally controlled air pressure, the capillary viscometer is rarely used for precise investigation of shear-rate dependence. Instead, concentric cylinder viscometers of the type shown in Figure 7.10(b) are most commonly employed, devices very similar to those already described for flow birefringence studies. One of the cylinders is driven at a constant speed, and the torque exerted on the other is measured. If the ratio of the gap between the cylinders to cylinder diameter is very small, the conditions approximate those of flow between infinite plates (Figure 7.1), and a nearly uniform rate of shear is obtained.

A simple and accurate version of this viscometer is shown in Figure 7.10(c). In this device the inner cylinder is a self-centering float, containing a small magnet. It is driven by an external revolving solenoid, and the viscosity of the liquid in the annular space causes the float to rotate with a constant velocity, determined by the dimensions of the apparatus and viscosity of the solution. This apparatus (called the Zimm viscometer) has become widely used in studies of high-molecular-weight DNA.

PROBLEMS

1. The following data describe the determination of the intrinsic viscosity of bovine serum albumin (BSA) in a solvent containing 33.3 percent dioxane and 0.03 M KCl at pH 2.0, $T = 25°C$. The protein concentration (C) is in grams per 100 ml; ρ/ρ_0 is the ratio of solution density to that of the solvent.

C	ρ/ρ_0	Flow Time Readings (sec)		
0.000	1.000	398.1,	398.2,	398.2
0.417	1.0011	439.3,	439.1,	439.1
0.685	1.0019	467.6,	467.6,	467.5
0.844	1.0023	485.8,	485.5,	485.7

(*a*) Calculate $[\eta]$.

(*b*) What would you conclude about BSA under these conditions? Compare with results in Table 7.1.

2. A spherical, unhydrated protein molecule has a radius of 25 Å and $\bar{v} = 0.740$. It dimerizes to form a dimer that is a prolate ellipsoid 100 Å long and of the same *volume* as two of the monomers. Calculate the percent change in s and $[\eta]$ upon dimerization and decide which method would be best for following the process.

3. According to P. Flory, the intrinsic viscosity of a random-coil polymer should be proportional to R_G^3/M, and the frictional coefficient should be proportional to R_G. From this, find a product of powers of s and $[\eta]$ that should provide a measure of M.

4. Prove that

$$\lim_{C \to 0} \frac{1}{C} \ln \eta_r = [\eta]$$

and hence that $[\eta]$ may be determined by plotting $(1/C) \ln \eta_r$ versus C.

5. What value of the shear gradient would be required to produce an extinction angle of 40 deg with tobacco mosaic virus? If the shear were produced in a concentric cylinder apparatus, in which the inner (stationary) cylinder had a radius of 5 cm and the outer (moving) cylinder a radius of 5.5 cm, approximately what radial velocity (revolutions per minute) would be required?

REFERENCES

General

Yang, J. T.: *Advan. Protein Chem.*, **16**, 323 (1961). A good general review of the theory and practice of viscometry.

Tanford, C.: *Physical Chemistry of Macromolecules*, John Wiley & Sons, Inc., New York, N.Y., 1961, Chap. 6.

Flow Dichroism

Wada, A.: *J. Polymer Sci.*, **A2**, 853 (1964). Apparatus and some experimental results.

Flow Birefringence

Cerf, R. and H. A. Scheraga: *Chem. Rev.*, **51**, 185 (1952). A general review.

Non-Newtonian Flow

Yang, J. T.: *J. Am. Chem. Soc.*, **80**, 1783 (1958). A nice experimental study.

Denatured Proteins as Random Coils

Tanford, C., K. Kawahara, and S. Lapanje: *J. Am. Chem. Soc.*, **89**, 729 (1967).

The Scheraga–Mandelkern β-Function

Scheraga, H. A. and L. Mandelkern: *J. Am. Chem. Soc.*, **75**, 179 (1953).

EIGHT | ABSORPTION AND EMISSION OF RADIATION

In the remaining chapters, we shall be dealing with interactions between radiation and molecules. As the reader has surely learned, electromagnetic radiation cannot be uniquely described on either a wave or particle basis; it must be considered to have some of the attributes of both. In the various ways in which matter and radiation interact, there are some that are most easily described by emphasizing the wavelike aspects (scattering, for example) and other processes, such as absorption, that compel us to consider the existence of photons. In what is to follow, we shall unashamedly use whichever point of view is most convenient at the moment, but the reader is cautioned that a most rigorous treatment of *all* the phenomena would be from a quantum-mechanical basis. We do not intend here to go deeply into either classic electromagnetic theory or quantum mechanics. Instead, we shall be concerned with the applications of scattering and absorption phenomena to the investigation of biological macromolecules.

8.1 ABSORPTION SPECTRA: GENERAL PRINCIPLES

In classical physics the absorption of radiation will occur when the frequency of the electromagnetic wave is close to or equal to a *natural vibration frequency* of the molecule. There are occasions when this analogy between mechanical phenomena and absorption will be useful, but in general it will not get us very far. This point of view cannot explain the existence of *sharp* spectral lines, nor does it relate the *amount* of energy absorbed by an atom or molecule to the frequency of the radiation.

The discussion of absorption processes, then, requires that we adopt the point of view of quantum mechanics. Basically, there are two ways in which this differs from the classical picture:

1. The requirement that the energy available for absorption be related to the frequency (v) of the radiation by Planck's law:

$$E = hv \qquad (8.1)$$

where h is Planck's constant (6.67×10^{-27} erg-sec). There is no equivalent to this in classical physics; it imposes, as it were, a particulate aspect on radiation.

2. Recognition of the fact that atoms and molecules can exist at only certain *energy levels*. These turn out to be natural consequences of a wave-mechanical picture of matter.

We are not concerned here with the details of how the allowed energy levels of a system are deduced. However, the results for two simple examples that throw light on much more complex systems will be presented.

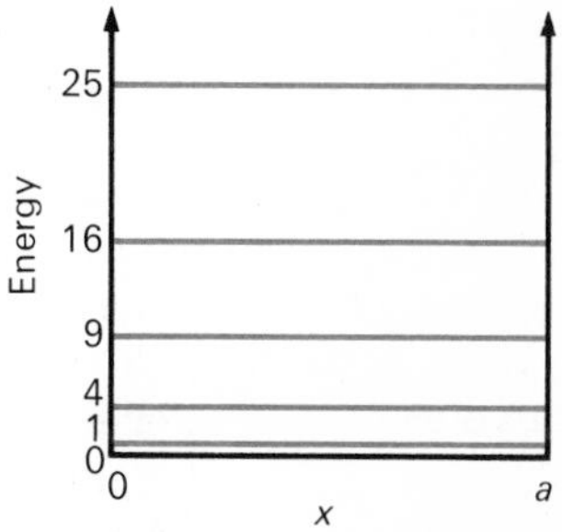

Figure 8.1

The simplest quantum-mechanical system—a particle in a one-dimensional box. The heavy line is potential energy versus distance; it is zero within the box ($0 \leq x \leq a$) and infinite outside ($x > a$ or $x < 0$). The fine lines show the allowed energy levels.

In classical physics a particle can have any energy. However, in quantum mechanics we find that constraining a particle automatically restricts its energy to only certain allowed levels. A specific, if somewhat abstract, example is shown in Figure 8.1. The particle (an electron, for example) is held in a one-dimensional "box" between $x = 0$ and $x = a$. Wave mechanics then requires that the particle can have only energies

$$E_n = n^2h^2/8ma^2$$

$$n = 1, 2, 3, 4, \ldots \qquad (8.2)$$

where m is the mass of the particle and n a quantum number. Absorption of radiation could then occur only for frequencies corresponding [by (8.1)] to transitions between these levels. Note that the spacing of the allowed levels is greater the smaller the box. We shall make use of this general principle later.

A more complex system (but still simple by biochemical standards) is a diatomic molecule. Here we must make distinctions between the different kinds of energy the system can have. In the first place the energy of the molecule will depend on the electronic state—specifically, the set of orbitals that the electrons in the molecule occupy. For a given state, the energy will depend on the distance between the nuclei, in a fashion similar to that shown by curves I or II in Figure 8.2. For each electronic state, our molecule will have a set of allowed levels of *vibrational* energy (the horizontal lines in Figure 8.2). Finally, the *rotational* energy is also quantized; these states would correspond to sets of even more closely spaced lines clustered above each vibrational level.†

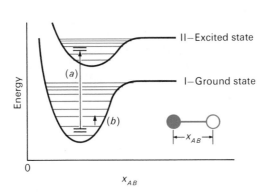

Figure 8.2

Energy levels for a simple diatomic molecule. The potential energy (heavy curve) is a function of the internuclear distance (X_{AB}). Two electronic states are shown (I and II). They differ in their potential energy and in the position of the minimum, the equilibrium internuclear distance. For each electronic state there are different possible levels of vibrational energy (long, thin lines) and rotational energy (short lines, shown only for two vibrational levels). A possible electronic transition is shown by (a) and a vibrational transition in the electronic ground state by (b).

These simple systems allow us to demonstrate the most important features of absorption spectroscopy. Transitions may occur between electronic, vibrational, or rotational levels. The energy change, and hence the frequency of the radiation involved, will generally decrease in the same order. Figure 8.3 schematically describes the electromagnetic spectrum and indicates the general ranges in which each type of transition is to be found. Note that the wavelength (λ) or frequency (v) can be expressed in a number of ways. For wavelength, one usually uses either angstrom units (Å), microns (μ), millimicrons (mμ), or nanometers (nm). Frequency is given in sec^{-1} (since $v = c/\lambda$), but sometimes the *wave number* $v' = 1/\lambda$ is given instead, usually in cm^{-1}.

Spectroscopy, to the biochemist or biophysicist, generally means the observation of absorption spectra. Except in the cases of fluorescence and phosphorescence, *emission* of electromagnetic radiation is appreciable only at temperatures too high to preserve the integrity of sensitive biological molecules. Most

---

† For a more detailed analysis of such quantum-mechanical problems, the student is referred to any of the better physical chemistry texts.

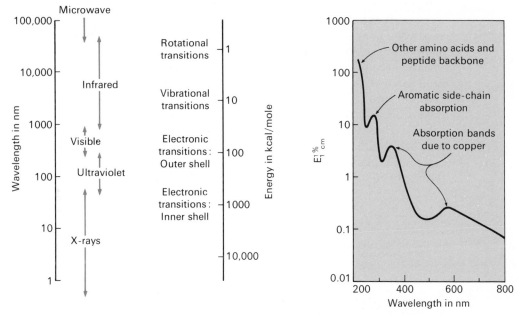

Figure 8.3   (*a*) A sketch of the electromagnetic spectrum. (*b*) The visible and near-UV spectrum of the copper protein, hemocyanin. Bands in the visible and near-UV arise from the copper moeity; those near 280 nm from aromatic protein side chains.

studies are made on solutions, and these are quite often dilute solutions. In such circumstances the simple Beer–Lambert law will govern the absorption processes:

$$I = I_0 e^{-\varepsilon' lc} = I_0 10^{-\varepsilon lc} \tag{8.3}$$

where I_0 is the intensity of the incident radiation and I the intensity of the radiation transmitted through a cell of thickness l cm, containing a solution of concentration c moles/L. The quantity ε is the *extinction coefficient*, with the units liters per mole per centimeter. Data are frequently reported in percent transmission ($I/I_0 \times 100$) or in absorbance [$A = \log(I_0/I)$]. The latter is particularly convenient, for from (8.3)

$$A = \varepsilon lc \tag{8.4}$$

Sometimes the extinction coefficient is given in other units; for example,

$$A = E_{1\,\text{cm}}^{\%} lC \tag{8.5}$$

where the concentration C is in grams per 100 ml of solution. This is useful when the molecular weight of the solute is unknown or uncertain.

The measurement of absorption spectra, then, requires that we be able to produce monochromatic light of known wavelength and measure the decrease in intensity that occurs when this light passes through a known thickness of solution. Today such intensity measurements are almost always made electronically, using either photocells or photomultiplier tubes. A sketch of a modern spectrophotometer is shown in Figure 8.4. The source (here a xenon arc) produces a wide range of radiation; the wavelength required is selected by a prism or grating monochromator. The beam is split, and the two portions traverse, respectively, the sample cell and a reference cell, containing only solvent. The intensities are measured by photomultiplier tubes, and the difference between the logarithms of sample and reference intensity is recorded. In many such instruments the wavelength scale is scanned automatically, with A recorded as a function of λ on a synchronized recorder.

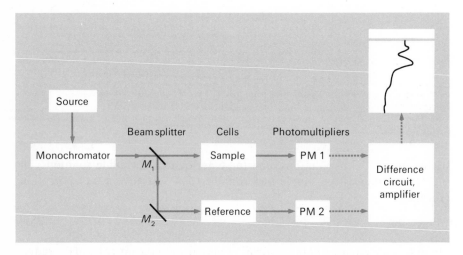

Figure 8.4   A simplified diagram of a recording spectrophotometer. See the text.

8.2 INFRARED SPECTRA

In Section 8.1 we sketched some general principles of spectroscopy. These ideas apply even to the largest molecules of interest to the biochemist, albeit with certain restrictions and complications.

In the first place, pure rotational spectra are rarely observed for biochemically interesting substances. Most of these materials are studied in solution; in this dense environment, pure rotational states cannot exist as they do in the gas.†

† While pure rotational spectra are not observed for molecules in solution, there exists low-frequency absorption that apparently corresponds to *librational* motions, which are rotational oscillations about equilibrium positions. This area of spectroscopy has been little explored.

Turning to the shorter wavelength region of the near infrared, we find an embarrassing wealth of detail in the absorption spectra of even the simpler molecules of biological interest. The reason for this is clear; here we are observing vibrational transitions and should be considering the possible normal modes of vibration of the molecules. A nonlinear molecule with n atoms will have $3n - 6$ fundamental modes of vibration. Even for a simple substance such as an amino acid, this yields a large number; for a macromolecule such as protein, a complete analysis of the infrared spectrum is quite beyond present imagining. Nevertheless, the fact that certain groups display vibrational transitions at characteristic frequencies makes some unscrambling possible. For example, the $-C=O$ group has a fundamental stretching frequency of about $1,700 \text{ cm}^{-1}$ in many molecules, and a fundamental stretching mode of the $-N-H$ group is almost always observed in the neighborhood of $3,400 \text{ cm}^{-1}$.

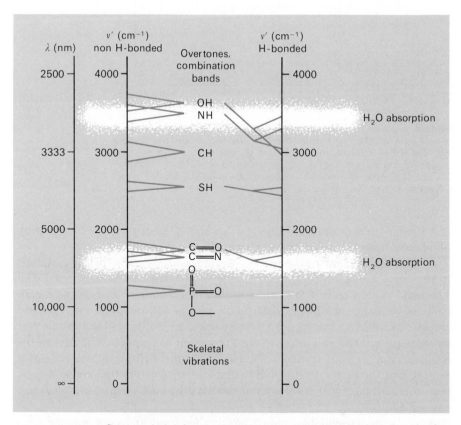

Figure 8.5 Some typical infrared absorption bands found in biological molecules. Approximate shifts on H-bonding are shown, and the regions of water absorption are shown within the white area.

A list of such frequencies is given in Figure 8.5. An example of the kind of information that can be obtained by the study of such identifiable infrared bands is shown in Figure 8.6. The polypeptide is in the α-helical configuration so that the —C=O groups are hydrogen-bonded. This results in a decrease of the stretching frequency from the value of about 1,700 cm^{-1} for free —C=O groups to a value (1,650 cm^{-1}) close to that observed in crystalline amides, where hydrogen bonding can be demonstrated by X-ray diffraction.

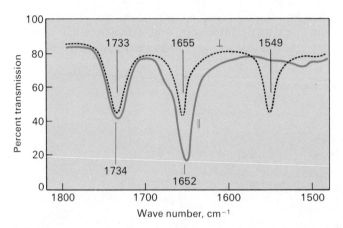

Figure 8.6   Infrared spectrum of a film of poly-γ-benzyl-L-glutamate in the α-helical form. A solid film of the polymer has been prepared and oriented by stretching. The dotted line is for radiation polarized perpendicular to the molecular orientation and the solid line for radiation polarized parallel to the helix axes. See the text. Adapted from M. Tsuboi, *J. Polymer Sci.*, **59**, 139 (1962). Copyright © 1962, John Wiley and Sons, Inc.

A similar study is shown in Figure 8.7, which demonstrates the base pairing of nucleoside derivatives in nonaqueous solution.

A great impediment to the use of infrared analysis of biological materials is the very strong absorption by water in the regions around 3,400 and 1,600 cm^{-1} (see Figure 8.5). This prohibits, for example, an otherwise easy study of hydrogen bonding of macromolecules in aqueous solution. To some extent, this difficulty can be overcome by using dried films of the substance or by working in D_2O solutions, where the solvent absorption is shifted to the less interesting regions around 1,200 and 2,500 cm^{-1}. However, one can never be sure that the subtle details of macromolecular structure that are so important in biochemical processes are not modified by such changes in molecular environment.

A good deal of useful information has resulted from studies of oriented films of elongated macromolecules using polarized infrared radiation. Under such circumstances, dichroism will be observed; the absorption will depend on the

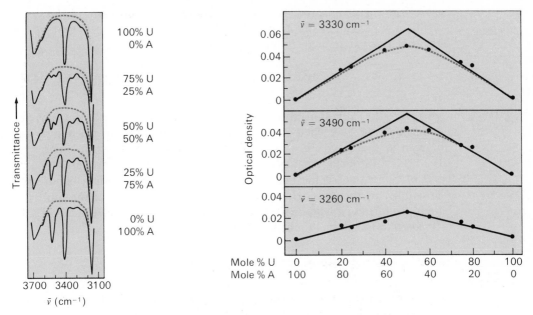

Figure 8.7 Infrared detection of the hyrdogen-bonded association of 1-cyclohexyl uracil and 9-ethyladenine in deuterochloroform solutions. *Left*, spectra (dotted lines are background) for different mixtures. *Right*, intensities of bands as a function of composition. The bands at 3,330, 3,490, and 3,260 cm^{-1} are indicative of hydrogen-bonding. From R. M. Hamlin, R. C. Lord, and A. Rich, *Science*, **148**, 1734 (1965). Copyright © 1965, by the American Association for the Advancement of Science.

relative orientation of the absorbing chromophoric group and the direction of polarization of the light. We can express this quantitatively by considering, as in Figure 8.8, that the *transition dipole moment* (very approximately, the charge displacement that accompanies the absorption) makes an angle θ with respect to the direction of light polarization. If the amplitude of the electromagnetic wave is represented by a vector **E** (see Chapter 9), the projection of this vector in the direction of the transition moment will be proportional to

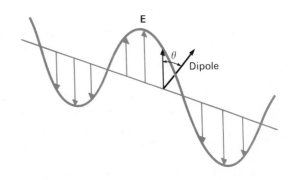

Figure 8.8

The basis of dichroism. The angle between the direction of polarization and the direction of the transition dipole determines the probability of absorption.

$\cos \theta$. The energy absorbed will be proportional to the intensity, or the *square* of this component of the amplitude. Thus the extinction coefficient, ε_θ, for light polarized at an angle θ to the transition moment will be given by

$$\frac{\varepsilon_\theta}{\varepsilon_\parallel} = \cos^2 \theta \qquad (8.6)$$

where $\varepsilon_\parallel$ is the extinction coefficient for light polarized parallel to the transition moment. Note that when $\theta = 90$ deg, $\varepsilon_\theta = 0$. This effect of orientation can be seen in the spectrum of poly-γ-benzyl-L-glutamate shown in Figure 8.6. The —C=O stretching bands ($\sim 1{,}650 \, \mathrm{cm}^{-1}$) are polarized strongly along the axis of elongation of the film, showing that the corresponding bonds must lie parallel to this axis. This is consistent with the idea that the molecules are in the form of α helices (see Barker, in this series) in which these bonds are supposed to lie parallel to the long axis of the molecules.

8.3   SPECTRA IN THE VISIBLE AND ULTRAVIOLET REGIONS

Transitions in the visible region of the spectrum are relatively low-energy *electronic* transitions. Roughly speaking, there are two main kinds of biological structures that have energy levels with spacings in this range: (1) Compounds containing certain metal ions (particularly transition metals) and (2) large aromatic structures and conjugated double-bond systems. The former include some metalloproteins and the latter such conjugated structures as vitamin A. Heme proteins owe most of their intense absorption in the visible region to the conjugated heme group rather than the metal ion. It will be noted that electronic absorption bands are usually quite broad. This is because they actually consist of a large number of closely spaced subbands, each corresponding to a different change in vibrational energy accompanying the change in electronic state (see Figure 8.2).

In the near-ultraviolet region of the spectrum (200 to 400 nm), many biochemicals exhibit strong and useful absorption bands. In keeping with the rough "particle in a box" rule, it is not surprising that these are often associated with *small* conjugated ring systems, both aromatic and heterocyclic, whereas the large conjugated systems absorb in the visible region. Thus among the amino acids, tyrosine, phenylalanine, and tryptophan absorb most intensely in this region (Table 8.1). Most of the absorption of proteins in the 280 nm region results from the presence of these residues. Similarly, all of the purine and pyrimidine bases absorb strongly in the near ultraviolet. Most of these absorption bands appear to correspond to transitions of π electrons in the ring to antibonding π orbitals (the so-called π–π^* transitions). Evidence for this is seen in Figure 8.9, where spectra of crystals of 1-methyl thymine are shown.

Farther in the ultraviolet, nearly *everything* absorbs. Suppose we consider the absorption spectrum of a hypothetical protein. The 280 nm absorption is due, of course, to the aromatic side chains, but below about 230 nm we begin to

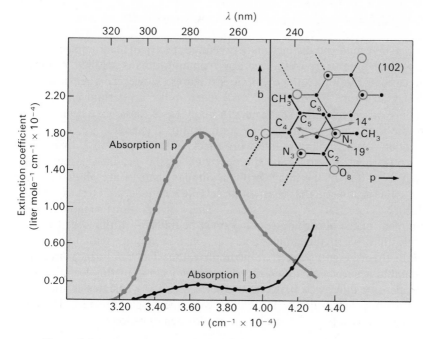

Figure 8.9 Polarized absorption spectra of crystals of 1-methylthymine. The absorption is from π–π* transitions; probable directions of transition moments are shown. From R. F. Stewart and N. Davidson, *J. Chem. Phys.*, **39**, 255 (1963).

TABLE 8.1 NEAR-ULTRAVIOLET ABSORPTION BANDS OF SOME AMINO ACIDS AND NUCLEOTIDES[a]

Substance	pH	λ_{max} (nm)	ε (liters/cm·mole)
Phenylalanine	6	257	200
Tyrosine	6	275	1,300
Trypotophan	6	280	5,000
Adenosine-5′-phosphate	7	259	15,000
Cytidine-5′-phosphate	7	271	9,000
Uridine-5′-phosphate	7	262	10,000
Guanosine-5′-phosphate	7	252	14,000

[a] In each case the longest wavelength band is given. Each compound has other (and generally stronger) bands in the vicinity of 200 nm.

encounter not only transitions in other amino acid side chains but also those involving electron displacements in the peptide backbone itself. There appears to be a particularly intense absorption of the latter type at about 200 nm and a weaker one at about 225 nm. Below about 180 nm lies *terra incognito*; work in this region is rendered exceedingly difficult by the absorption by oxygen of the air, in addition to that by water and almost all other solvents. Experiments seem possible only with thoroughly dried materials in evacuated spectrographs—hence the term *vacuum ultraviolet*.

8.4   MACROMOLECULAR STRUCTURE AND ABSORPTION

So far, we have made no distinction between the spectra of macromolecules and those of their monomers. As a first approximation this is justified, especially in the case of random-coil polymeric structures. However, when a definite secondary structure is established in a macromolecule, certain changes are commonly observed in the spectrum. In many cases these have been used as diagnostic tools to detect and follow configurational changes.

Consider, for example, the following effect: When a protein is hydrolyzed to its constituent amino acids, the spectrum of the products is somewhat different from that of the intact protein. Similar changes are observed when the protein is *denatured*, that is, when its definite secondary and tertiary folding are disturbed (see Barker, in this series). These observations suggest that the environment of the absorbing residues must be different in the folded protein molecule than when the amino acids are in free contact with the solvent. While the theory is still somewhat incomplete, the effect can be measured with considerable accuracy by *difference spectrophotometry*. In this technique the optical density difference between the substance in two states is measured directly (using, for example, a double-beam spectrophotometer). Thus the change can be used to follow such processes as denaturation.

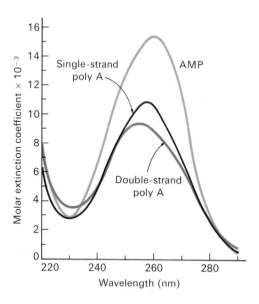

Figure 8.10

Hypochromism of polyriboadenylic acid. Note that the form of the polymer spectrum differs little from that of the monomer but that the intensity of absorption is considerably reduced.

With nucleic acids even more dramatic effects are observed. Figure 8.10 contrasts polyriboadenylic acid with an equal concentration of its monomer, adenosine monophosphate, while Figure 8.11 shows the change produced in the spectrum of *E. coli* DNA by heating to 90°C with accompanying denatura-

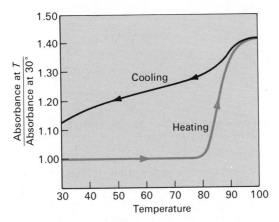

Figure 8.11 The "melting" of DNA as followed by absorption at 260 nm. Note the sharpness of the transition in cycle 1. This is typical of a cooperative process (see Chapter 3). Also note that recovery of the ordered structure is not complete on rapid cooling.

tion.† Let us consider the simpler case of the synthetic polyriboadenylic acid. Two effects are observed; the polymer shows a *lower* absorption intensity than the monomer (hypochromism), and in the polymer the maximum is shifted to a wavelength almost 3 nm lower.

Both these effects have been explained on the basis of the helical configuration of the nucleotide polymers (see Wold, in this series). In such a regular, closely packed structure the chromophores cannot be considered to be independent of one another. The electronic displacement that occurs when one base absorbs a quantum of energy is felt in the neighboring bases as a modification of the electrical field. If the interaction is strong, it becomes impossible to speak any longer of an absorption "localized" in a particular residue. Rather than the single, N-fold degenerate absorption frequency that a polymer of N independent groups would exhibit, an *exciton band*, with N closely spaced but distinct levels, is now observed. The total absorption intensity will not necessarily be distributed equally among the various levels; in this way a shift in the maximum wavelength of absorption may be produced.‡

A simple example will illustrate the principle. Suppose we have a dimer of a substance that in the monomeric form has an absorption at the frequency v_0. If in the dimer there is interaction of energy U between the chromophores, the single monomer band will be split into two bands at frequencies

$$v_1 = v_0 + U/h, \qquad v_2 = v_0 - U/h \tag{8.7}$$

† For a detailed discussion of the thermal denaturation of nucleic acids, the reader is referred to Barker, in this series.

‡ Of course, this is not the only way in which a change in absorption frequency can arise. Simply putting a chromophore into a different environment (inside a helix, for example) may change its electronic properties.

 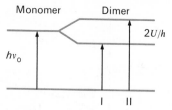

Figure 8.12   Splitting of a transition upon dimer formation. The two modes of excitation in the dimer lead to different excited-state energies.

(see Figure 8.12). The relative absorption intensities will depend on the geometric relation between the transition moments in the two chromophores. In a real system the vibrational broadening of the bands would probably prevent resolution, and we might observe only a shift in the frequency of maximum absorption.

The hypochromism of polymeric structures also arises from intramolecular interactions. Due to interactions with the solvent and surrounding polymer, some of the transitions of the monomer groups may be more favored and others less. Thus the hypochromism of the 260 nm band in polyadenylic acid or DNA, and its change with conformation, is not surprising. A very general rule of spectroscopy holds that mere configurational modifications that do not change the fundamental electronic structure should not change the *total* absorption, integrated over all bands. If this is to hold for the nucleic acids in question, we must postulate an equal *hyperchromism* somewhere else, presumably in the inaccessible low-wavelength region.

The hypochromism in the ultraviolet bands of the nucleic acids has become an exceedingly useful tool for the observation of configurational change (see Wold, in this series), since spectroscopic absorption measurements are rapid and precise.

8.5   FLUORESCENCE AND PHOSPHORESCENCE

In discussing absorption we have been concerned entirely with the excitation of a molecule from a ground state to a higher energy level and have given no consideration to the subsequent fate of the excited molecule. In many cases the sequel is not very interesting; the energy is transferred, as heat, to the surroundings. However, in some cases reradiation occurs and this will usually be at a different frequency than that of the exciting light. This process is called *fluorescence*.

Figure 8.13(*a*) shows, in a crude fashion, how a substance can exhibit fluorescence. If the molecule is in the lowest vibrational level of the electronic ground state, its internuclear distance will most likely be about that shown by point (*a*) in the figure. This is dictated by the form of the vibrational wave function, which will give a maximum probability for this internuclear distance. If an electronic excitation occurs, it will happen so quickly that no appreciable

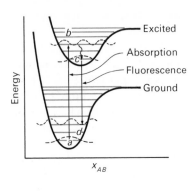

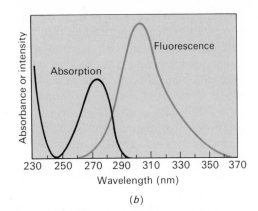

(a)   (b)

Figure 8.13   (a) Fluorescence arising from absorption by a bonding electron. The figure is similar to Figure 8.2, except that the forms of vibrational wave functions are shown (dotted lines). The height of the dotted line above its limiting values indicates the probability of different internuclear distances. (b) The excitation and emission spectra for the fluorescence of tyrosine.

motion of the massive and sluggish nuclei will have taken place (the Franck–Condon principle). Thus we should expect the transition to be described by a nearly vertical line (*ab* in the figure), and the most likely transitions will be those that lead to vibrational states that have maximum probability of internuclear distance close to the initial distance. In the example shown, the form of the potential energy curve for the excited stated is such that the third vibrational level in this state will be favored.† Two things may now happen: (1) all of the excitation energy may be lost in a nonradiative manner, or (2) the molecule may lose some energy as heat and thus reach the lowest vibrational level of the excited state and then reradiate (line *cd*). If this happens, the emitted light will be of lower frequency (and thus longer wavelength) than the exciting light, and fluorescence will be observed.

It is commonly found that the fluorescence spectrum is quite independent of the wavelength of the exciting light. This can be explained by the fact that the time required for the loss of excess vibrational energy in the excited state (about 10^{-12} sec) is very much less than the lifetime of the excited electronic state (about 10^{-9} sec). Thus excited molecules have plenty of time to reach the lowest vibrational level before they re-emit, and the form of the fluorescence spectrum will be dictated by the relative probabilities of falling to the various vibrational levels in the *ground state*. Furthermore, transfer of electronic excitation within a given chromophore is a rapid process; the result is usually that excitation to any electronic level is followed by fluorescence from the *lowest* excited electronic state. The fluorescence spectrum thus serves, like the absorption spectrum, as a specific fingerprint for a compound. This, together with the fact that very small amounts of *emitted* radiation can be detected, makes

† For a fuller understanding of this point, some knowledge of the forms of the vibrational wave function is needed. See, for example, N. J. Moore, *Physical Chemistry*, 3rd edition (Prentice-Hall, Inc., Englewood Cliffs, N.J., 1962).

fluorescence an exceedingly sensitive analytical tool. Figure 8.13(*b*) shows the typical fluorescence spectrum of tyrosine.

If the exciting light is polarized, absorption will be most likely for those molecules that happen to have their chromophores oriented with transition moments parallel to the direction of polarization [see Equation (8.6)]. If, in the 10^{-9} sec or so between absorption and emission the molecules do not rotate appreciably, the emitted light will also be polarized. However, because of rotational Brownian motion (see Chapter 6), some *depolarization* will generally occur. This is a convenient situation, for it means that we can use this depolarization to measure the rate of Brownian rotation, which will depend on the dimensions (and particularly the asymmetry) of the molecule carrying the chromophoric groups.

Recently, it has become possible to make this measurement in a direct way by using light pulses of very short duration (about 10^{-9} sec). If the fluorescent emission following such a pulse is measured with detectors fitted with polarizers, the intensity of emission polarized parallel to ($I_\parallel$) and perpendicular to ($I_\perp$) the incident pulse can be compared. The emission anisotropy (A) is defined as

$$A = (I_\parallel - I_\perp)/(I_\parallel + 2I_\perp) \tag{8.8}$$

The anisotropy decays exponentially with time because the molecules that emit later have had more time to rotate:

$$A(t) = A_0 e^{-3t/\rho} \tag{8.9}$$

where ρ is the rotational relaxation time (compare with the expressions for the decay in electrical dichroism and birefringence, Chapter 6). Since very short relaxation times (of the order of 10^{-9} sec) can be measured in this way, a new and powerful probe for changes in molecular structure has become available.

There is yet another way to measure rotational relaxation from fluorescence depolarization, which requires simpler apparatus but involves more difficulty in interpretation. An apparatus for such experiments is sketched in Figure 8.14. If the sample is irradiated continuously with exciting light, a *steady-state* polarization of the fluorescence will be observed, which will depend on the ratio of the average excited-state lifetime (τ) to the rotational relaxation time (ρ). Such measurements have been traditionally expressed in terms of a quantity

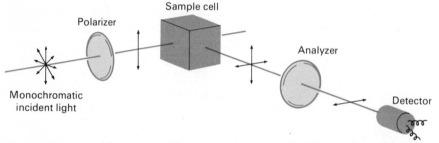

Figure 8.14   Measurement of fluorescence depolarization. The fluorescent light is detected at 90 deg through an analyzer that rejects all but the depolarized component.

called the *polarization*:

$$P = \frac{I_{\parallel} - I_{\perp}}{I_{\parallel} + I_{\perp}} \tag{8.10}$$

In these terms one finds

$$\left(\frac{1}{P} \pm \frac{1}{3}\right) = \left(\frac{1}{P_0} \pm \frac{1}{3}\right)\left(1 + \frac{3\tau}{\rho}\right) \tag{8.11}$$

where the plus and minus signs correspond to unpolarized and polarized incident radiation and P_0 is the intrinsic polarization of the fluorescence (the polarization that would be observed if no molecular rotation occurred). Equation (8.11) is not especially useful as it stands, however, since neither τ nor P_0 are usually known. If the rotational relaxation time is approximated[†] by $\rho \simeq 3\eta V/RT$, where η is the solvent viscosity and V an effective molar volume, then

$$\left(\frac{1}{P} \pm \frac{1}{3}\right) = \left(\frac{1}{P_0} \pm \frac{1}{3}\right)\left(1 + \frac{RT}{V\eta}\tau\right) \tag{8.12}$$

If measurements are made in solutions of varying viscosity and/or temperature, the left-hand side of Equation (8.12) can be graphed versus T/η. Extrapolation to $T/\eta \to 0$ gives P_0; assumption of a value of τ will then give ρ from Equation (8.10). This procedure encounters one difficulty; to vary η one must either change the temperature of the solution or add a third component (such as sucrose or glycerol). In Figure 8.15 are shown graphs of the fluorescence

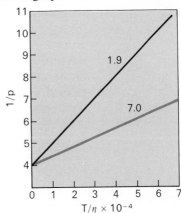

Figure 8.15
Depolarization of the fluorescence of bovine serum albumin. Data are given at pH 1.9 and 7. Note that the relaxation time is less at the lower pH. Adapted from G. Weber (1952).

---

† This results from the fact that the rotational relaxation *time* is inversely proportional to the rotational diffusion coefficient (Chapter 6), which measures the *rate* of rotational disorientation. Since $\rho = 1/2\Theta$ for spherical molecules, we have in this case, from Equation (6.21),

$$\rho = \frac{4\pi \mathcal{N}\eta R^3}{RT} = \frac{3\eta V}{RT}$$

polarization of bovine serum albumin in neutral and acidic solution. The fact that the rotation relaxation time is *lower* in acidic solution is surprising, for sedimentation and viscosity data have been interpreted in terms of an *expansion* of the molecule in acidic solution. The fluorescence depolarization results show that the process is more complex than simple expansion, which would lead to a *longer* relaxation time. One interpretation holds that the serum albumin molecule consists of several loosely but covalently connected subunits that can move apart and rotate independently at low pH.

In our discussion of fluorescence so far, we have assumed that the chromophore that emitted the fluorescent radiation was the same group as the absorber. This need not be the case; under favorable circumstances excitation energy can be transferred from one chromophore to another. The requirements are (1) the possibility of dipole–dipole interaction between the chromophores and (2) an appreciable overlap of the fluorescence spectrum of the donor and the absorption spectrum of the acceptor. The fact that the transfer appears to operate by a dipole interaction mechanism explains the strong dependence of such transfer on the distance between the participating groups. The interaction is found to vary with the inverse sixth power of the separation r. The efficiency of transfer is given by

$$\text{Efficiency} = \frac{1}{1 + (r/R_0)^6} \tag{8.13}$$

Here R_0 is a distance parameter characteristic of the donor–acceptor pair and the medium between them. According to Equation (8.13), the efficiency increases from very small values to near unity if r becomes less than R_0. Since R_0 is ordinarily found to be of the order of 20 Å, the observation of such sensitized fluorescence thus serves as a useful yardstick for the distances between groups in macromolecules. The fact that the phenylalanine, tyrosine, and tryptophan groups in proteins usually satisfy both requirements for transfer (proximity and spectral overlap) explains an otherwise mystifying observation. Even if excitation is in the phenylalanine or tyrosine bands, the observed fluorescence is usually from tryptophan. This is because the excitation energy is transferred to tryptophan, which has the lowest-lying excited state (longest wavelength absorption) of the three. Thus energy transfer should be thought of as a common rather than as an unusual phenomenon.

Another mode of de-excitation, called *phosphorescence*, deserves mention. In phosphorescence, the molecule transfers to a long-lived excited state. Such states are *triplet* states, that is, electronic states in which the two electrons in an orbital have their spins in the same direction. Transitions between triplet states and the normally singlet (paired-electron) ground states are "forbidden" by quantum-mechanical selection rules. In practice this means that the rate of dropping back to the ground state is very low, thus accounting for the long life of the excited state. Of course, at room temperature radiationless transfer is much more probable than phosphorescent emission, so that phosphorescence is most commonly observed at very low temperatures.

Since singlet-triplet transitions are strongly forbidden as well, we must ask how the triplet state is reached in the first place. Two mechanisms are common: (1) the molecule may be raised to a singlet excited state in the ordinary fashion and then cross over to a triplet state of about the same energy, or (2) a chemical reaction may leave the products in such an electronic state.

8.6   MAGNETIC RESONANCE METHODS

There are a pair of techniques, *electron paramagnetic resonance* (epr) and *nuclear magnetic resonance* (nmr), that have very recently become of considerable importance in the study of biological materials and processes. Both of these are actually spectroscopic methods, and they have many features in common, but their applications are rather different. Both depend on the following facts: A spinning charged particle behaves as a magnet; it possesses a *magnetic dipole moment*, which can be depicted as a vector along the axis of spin. Quantum mechanically, the spin a particle such as an electron or nucleus may have is quantized; the moment may have only certain values, measured in terms of a spin quantum number. If such a particle has spin other than zero, it will interact with an externally imposed magnetic field. By the rules of quantum mechanics, this interaction is quantized; the spin vector can take only certain allowed orientations with respect to the field (see Figure 8.16).

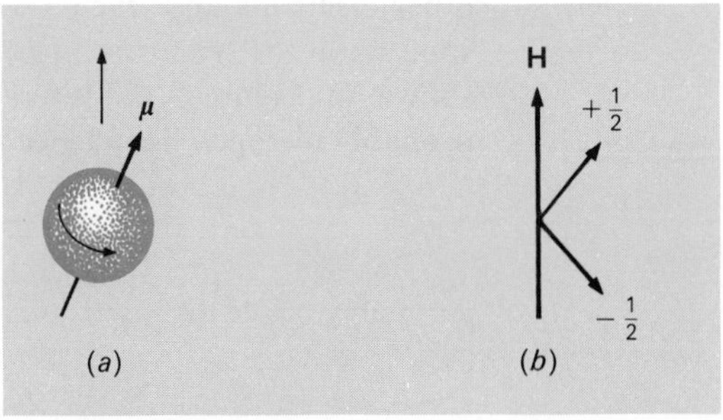

Figure 8.16

(*a*) A spinning charge of magnetic moment μ in a magnetic field H. (*b*) The two quantum mechanically allowed orientations of a particle of spin quantum number $S = \frac{1}{2}$.

These orientations correspond to different energy levels for the particle in the field. These energy levels are rather closely spaced, so that transitions between them correspond to radiation in the very far infrared, the microwave region. Furthermore, since the spacing depends on the external field, it is possible to keep the microwave radiation at a constant frequency and sweep the magnetic field so as to bring different transitions into resonance.

With this general introduction, let us consider the two techniques individually.

Electron Paramagnetic Resonance

Electron paramagnetic resonance depends on the existence of *unpaired* electrons in the sample. In general, there are two common sources for such electrons:

(1) transition metal ions with unpaired d electrons and (2) free radicals. Investigation of free-radical species has been of importance in the study of some biochemical mechanisms, but the principal use of the epr method has been in the study of metalloproteins which contain transition metal ions.

The basic concept can be described in terms of the behavior of an unpaired electron, assumed not to be magnetically interacting with other electrons or nuclei. Such an isolated electron will, in an external field, H, have two allowed orientations of its spin vector. These correspond to energies of nearly $+\mu_0 H$ and $-\mu_0 H$, where μ_0 is the *Bohr magneton*:

$$\mu_0 = eh/4\pi mc \tag{8.14}$$

Here e is the charge, m is the mass of the electron, h is Planck's constant, and c is the velocity of light. A transition between the two levels should correspond to the absorption of energy given by $h\nu = 2\mu_0 H$; exactly, it is found that

$$\nu = 2.0023\mu_0 H/h \tag{8.15}$$

The factor 2.0023 is termed the g *value*. The value given is for an isolated electron. In other circumstances the g value will differ very slightly from this number.

The above scheme predicts a single line for the epr spectrum. In practice the situation is always complicated by magnetic interaction of the electron with other electrons and nuclei. The latter type of interaction leads to *hyperfine* splitting of the spectrum. The epr spectrum of copper–conalbumin complex is shown in Figure 8.17. Interaction of the unpaired electron with what are probably nitrogen ligands leads to the observed splitting of the main peak.

Figure 8.17

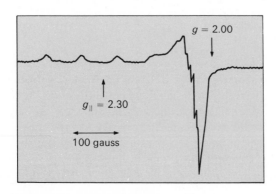

$g = 2.00$

$g_{\parallel} = 2.30$

100 gauss

The epr spectrum of a copper-conalbumin complex. As is usually the case, this has been recorded as a derivative spectrum. Two kinds of splittings can be seen. To the left (low field) are peaks due to interaction of the d electron with the spinning Cu nucleus. In the larger peak to the right are hyperfine splittings resulting from interactions with N ligands. Taken from J. J. Windle, A. K. Wiersema, J. R. Clark, and R. E. Feeney, *Biochemistry*, **2**, 1341 (1963). Copyright © 1963 by the American Chemical Society.

Nuclear Magnetic Resonance

Nuclear magnetic resonance is confined to the study of those nuclei that have nonzero spin. The spin quantum numbers, I, can take half or integral values. Nuclei that have been investigated include (spin numbers in parentheses) $^1H(\frac{1}{2})$, $^2H(1)$, $^{13}C(\frac{1}{2})$, $^{14}N(1)$, $^{17}O(\frac{5}{2})$, and $^{31}P(\frac{1}{2})$. Most work has been done with proton resonance. Most of the other usable nuclei are not the naturally abundant ones, and artificial enrichment is required to yield detectable signals.

The allowed energy levels for a nucleus of spin number I in a magnetic field of intensity H are given by

$$E_M = M\mu H/I \tag{8.16}$$

where μ is the nuclear magnetic moment and M a quantum number that can take the values $I, I - 1, \ldots, 0, \ldots - (I - 1), -I$. For a nucleus with $I = \frac{1}{2}$, two states are possible: $E = \pm\mu H$, and thus $\Delta E = 2\mu H$ (see Figure 8.16). If the microwave probe frequency is v, then the field required for resonance is

$$H = hv/2\mu \tag{8.17}$$

If the situation were as simple as this, nmr would be an interesting but not especially useful phenomenon. However, the interaction of the nucleus with its surroundings gives rise to a number of important effects. In the first place, the magnetic field that is experienced by a given nucleus depends not only on the externally imposed field but on the fields produced by electrons and other nuclei in the molecule. Thus, for example, protons in different parts of a molecule, or under different molecular conformations, experience different total fields and require a slightly different external field to reach resonance. These *chemical shifts* can be measured with great precision and provide a sensitive probe of nuclear environment. An example may make this clear. The proton resonance spectrum of purine in aqueous solution is shown in Figure 8.18. Note that the different kinds of protons exhibit different resonance fields and

Figure 8.18

The nmr spectrum of purine in 0.8 M aqueous solution. The four lines from left to right are (1) CH_3Cl protons (an external reference), (2) H_8, (3) H_2, and (4) H_6. The N-9 proton is not observed due to exchange. From M. P. Schweitzen, S. I. Chan, G. K. Helmkamp, and P. O. P. Tso, *J. Am. Chem. Soc.*, **86**, 696 (1964). Copyright © 1964 by the American Chemical Society.

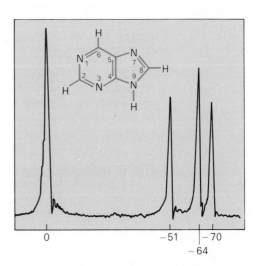

may be distinguished. In concentrated solutions, purine molecules associate by the same kind of stacking interaction that helps stabilize polynucleotide helices. This sandwichlike juxtaposition of molecules gives rise to different chemical shifts in different protons (see Figure 8.19). It was possible to decide that hydrogen bonding (which should yield an opposite shift for some protons) is *not* involved in these interactions.

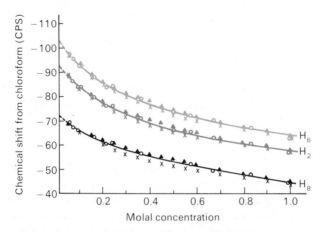

Figure 8.19 The concentration dependence of the chemical shifts observed in Figure 8.18. These results are consistent with a stacking of the purine molecules in aqueous solution: ○, experimental values; × and ▲, calculated values from two models of the stacking process. See S. I. Chan, M. P. Schweizer, P. O. P. Tso, and G. K. Helmkamp, *J. Am. Chem. Soc.*, **86**, 4182 (1964). Copyright © 1964 by the American Chemical Society.

A further complication in nmr spectroscopy that provides additional structural information is *spin coupling*. Just as in epr we observe interaction between the magnetic moment of a spinning electron and other magnetic moments in the environment, so there will also be interaction between the moments of proton spin. Most *inter*molecular interactions will be averaged out by molecular tumbling in gases and liquids, but spins may also be coupled through *intra*-molecular interactions, and such interaction will not average out. For example, protons attached to the same or adjacent carbon atoms may have quite strongly interacting spins. Since the energy of a particular spin state of a particular proton will then depend on the spin states of other protons, a splitting of the nmr lines will be observed. The magnitude of the splitting will depend on the strength of the interaction, making the assignment of lines easier, and will provide information about molecular geometry.

The nmr spectra of macromolecules are very complex, as might be expected. Only recently, with the advent of instruments with very high magnetic fields, has it been possible to obtain the kind of resolution needed to analyze such spectra. It is to be expected that nmr, utilizing both protons and other nuclei, will become a very important tool for the study of macromolecular structure in solution.

PROBLEMS

1. Cytosine has a molar extinction coefficient of 6×10^3 at 270 nm at pH 7. Calculate the absorbance and percent transmission of 1×10^{-4} and 1×10^{-3} M cytosine solutions in 1-cm and 1-mm cells.

2. A protein with extinction coefficient $E_{1\,cm}^{1\%} = 16$ yields an absorbance of 0.73 when measured in a 0.5-cm cell. Calculate the weight concentration.

3. Calculate, in kilocalories per mole, the energies corresponding to wavelengths of 10,000 nm (infrared), 250 nm (ultraviolet), and 1 nm (X ray).

4. The exciton splitting in polynucleotides appears to be of the order of 5 nm, with the absorption bands lying near 250 nm. Calculate, in kilocalories per mole, the corresponding interaction energies.

5. Flourescence depolarization studies of ovalbumin have been reported by G. Weber [*Biochem. J.*, **51**, 155 (1952)]. The protein was "tagged" with a dye that had a fluorescent lifetime (τ) of about 1.4×10^{-8} sec. The following data were recorded:

$T(°C)$	P
4.0	0.18
14.8	0.163
25.9	0.148
37.5	0.131
48.5	0.117

Determine the mean rotational relaxation time and compare it with the value estimated for a spherical molecule of the weight ($M = 45,000$) and specific volume (0.75) of ovalbumin.

REFERENCES

Spectroscopy of Macromolecules

Weber, G.: "The Interaction of Proteins with Radiation," in *The Proteins* (H. Neurath, ed.), 2nd ed., Vol. III, Academic Press, Inc., New York, 1966.

Wetlaufer, D. B.: *Advan. Protein Chem.*, **17**, 303 (1962). Weber's reference is more general, whereas Wetlaufer gives much detail on proteins and amino acids.

Fluorescence Depolarization

Stryer, L.: *Science*, **162**, 526 (1968). A short review of some of the newest techniques and results.

Weber, G.: *Biochem. J.*, **51**, 145, 155 (1952). These two pioneering articles give the theoretical basis and some early experimental work.

epr Spectrometry

Beinert, H. and G. Palmer: *Advan. Enzymol.*, **27**, 105 (1965). A broad discussion of epr application to biochemistry by two experts.

nmr Spectrometry

Roberts, J. D.: *Nuclear Magnetic Resonance*, McGraw-Hill Book Company, Inc., New York, 1959. A short book on the application of nmr to chemical structural problems.

NINE | SCATTERING

Absorption is not the only manner in which molecules can interact with radiation. Suppose we illuminate a solution with light of a wavelength that is far from any absorption band. Molecules are polarizable; their distribution of electronic charge can be shifted by an electric field. The classic description pictures radiation as a sinusoidally varying electromagnetic field. This should produce a sinusoidal oscillation of the electrons within a molecule. *Such oscillating charges will cause each molecule to act as a minute antenna, dispersing some of the energy in directions other than the direction of the incident radiation.* This is the basis for all scattering phenomena.

9.1 LIGHT SCATTERING; FUNDAMENTAL CONCEPTS

Since Chapter 9 through 11 will be concerned primarily with the various consequences of scattering, we should look at the process in detail. Consider the simplest situation, the scattering from a single molecule. We shall make the light monochromatic (by filters, for example) and polarized, although we shall remove the latter restriction later. The light is directed along the x axis in Figure 9.1 and polarized with electric vector in the z direction. The molecule is at $x = 0$, $y = 0$, and $z = 0$. We assume λ, the wavelength, to be so long that we can put the molecule *at* the origin, without having to worry about its size. Later, we shall consider what happens if there are many molecules in the scattering sample, or if the molecule is very large, or if the wavelength of the radiation is very small.

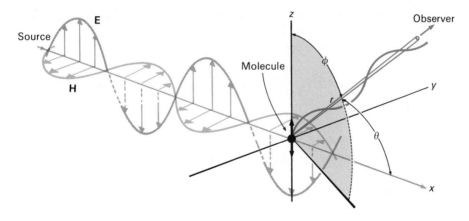

Figure 9.1   Scattering of radiation by a particle. Only the electric vector of the radiation scattered in the particular direction given by θ and ϕ is shown. Radiation scattered in this direction is polarized in the plane defined by the Z axis and r.

Since absorption phenomena are not to be considered, a classical rather than a quantum–mechanical treatment will be appropriate. The incident radiation can be described in terms of the electric field

$$\mathbf{E} = \mathbf{E}_0 \cos 2\pi v \left(t - \frac{x}{c} \right) \tag{9.1}$$

and the perpendicular magnetic field

$$\mathbf{H} = \mathbf{H}_0 \cos 2\pi v \left(t - \frac{x}{c} \right) \tag{9.2}$$

For the present discussion, only the electric field is of importance.† The electric field at the molecule ($x = 0$) varies with time as

$$\mathbf{E} = \mathbf{E}_0 \cos 2\pi v t \tag{9.3}$$

Since all molecules are polarizable, this field will produce an oscillating dipole moment

$$\boldsymbol{\mu} = \alpha \mathbf{E} = \alpha \mathbf{E}_0 \cos 2\pi v t \tag{9.4}$$

where α is the molecular polarizability. If the molecule is isotropic, the dipole moment will be in the direction of the electric vector, that is, along the z axis. This oscillating moment will act as a source of radiation; we may define the radiated field in terms of the coordinates shown in Figure 9.1. The amplitude

† The magnetic field and its interactions with electrons will be important in the discussion of optical rotatory phenomena. See Chapter 10.

of the electric field produced by an oscillating dipole, at a distance r from the dipole and at an angle ϕ with respect to the direction of polarization (the z axis), is given by electromagnetic theory† :

$$E_r = \left[\frac{\alpha E_0 4\pi^2 \sin \phi}{r\lambda^2} \right] \cos 2\pi v \left(t - \frac{r}{c} \right) \tag{9.5}$$

The term in brackets in Equation (9.5) is the part of interest to us. It represents the amplitude of the scattered wave. The intensity of the radiation (the energy flow per square centimeter) depends on the square of the amplitude. We wish to compare the intensity i of the scattered radiation to the intensity I_0 of the incident radiation. The latter is proportional to the square of *its* amplitude, E_0 :

$$\frac{i}{I_0} = \frac{(\alpha E_0 4\pi^2 \sin \phi / r\lambda^2)^2}{E_0^2} = \frac{16\pi^4 \alpha^2 \sin^2 \phi}{r^2 \lambda^4} \tag{9.6}$$

This equation tells a great deal about the scattered light. Its intensity falls off with r^{-2}, as radiation from a point source must. The intensity of scattering increases rapidly with decreasing wavelength.‡ The intensity depends on the angle ϕ; there is no radiation along the direction in which the dipole vibrates. A graph, in polar coordinates, of the radiation intensity looks like the doughnut-like surface in Figure 9.2(a).

Normally, unpolarized radiation is used in light-scattering experiments. Since this may be regarded as a superposition of many independent waves, polarized in random directions in the yz plane, we may imagine that the resulting scattering surface would correspond to the addition of surfaces such as that shown in Figure 9.2(a), rotated at random with respect to one another. The resulting surface is seen in Figure 9.2(b); it looks somewhat like a dumbbell, with the narrowest part in the yz plane. The equation for scattering of unpolarized radiation can be easily obtained‡‡; it differs from Equation (9.6) only in that $(1 + \cos^2 \theta)/2$ is substituted for $\sin^2 \phi$:

$$\frac{i}{I_0} = \frac{8\pi^4 \alpha^2}{r^2 \lambda^4} (1 + \cos^2 \theta) \tag{9.7}$$

Here θ is the angle between the incident beam and the direction of observation. The distribution of scattering intensity in the xy plane is shown in Figure 9.2(c); evidently, the scattering is symmetrical in forward and backward directions.

† The electromagnetic field surrounding a moving accelerating charge [the displaced electron(s)] will consist of three parts: the static field, depending only on the magnitude of the charge; the induction field, depending on its velocity; and the radiation field, depending on the acceleration of the charge. Only the last concerns us here, for the first two fall off rapidly with distance from the charge. See R. P. Feynman et al. (1963).

‡ The strong dependence of scattering on λ accounts for the blue of the sky, since we observe the sunlight scattered by the air and its contaminants, and the blue light is scattered more.

‡‡ See C. Tanford (1961).

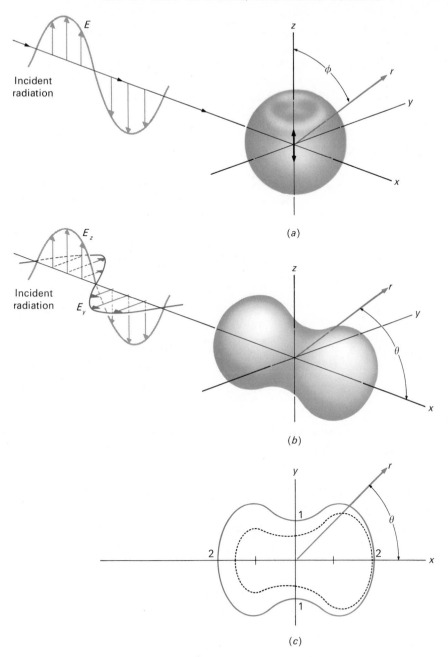

Figure 9.2   (*a*) Distribution of intensity of scattering of incident light polarized in the Z direction. In this figure, and in (*b*), the distance from the origin to the surface along a direction r shows the intensity in that direction. (*b*) Distribution of scattering intensity for unpolarized incident light. (*c*) A section in the xy plane of the surface of (*b*). The solid line shows the intensity of scattering as a function of θ for Rayleigh scattering (small particle). The dotted line is for a larger particle.

Equation (9.7) provides a completely adequate description of the scattering of light by a single small, isotropic particle. However, as it stands, Equation (9.7) is not of much use to the biochemist, who wishes to study solutions and to use the measurement of scattering to find out something about the solute molecules. We must therefore consider the situations that arise when the scattering is by a collection of particles.

9.2   SCATTERING FROM A NUMBER OF PARTICLES

To see the kind of thing that happens when scattering by more than one particle is occurring, let us consider the simplest case: two particles. Imagine, as in Figure 9.3, two identical scattering centers, fixed in space and illuminated by

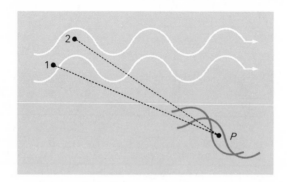

Figure 9.3

Scattering from a pair of atoms or molecules. The point of observation, *P*, is assumed to be far from the scatterers.

a common source. The passing electromagnetic waves stimulate each center to scatter. We observe the scattering at some distant point *P*. What is the amplitude and intensity of the scattered radiation at *P*? In general, there will be a phase difference in the scattered waves received at *P* from 1 and 2. There will be two reasons for this: First, the centers 1 and 2 lie, at any instant, at different positions in the incident field—they will scatter out of phase. A further phase difference can result from the different distances of *P* from points 1 and 2. To keep the expressions as simple as possible, we can represent the electric fields seen at *P* by the expression

$$A \cos(2\pi vt + \phi_1) \qquad \text{from center 1}$$

and $\hspace{10cm}$ (9.8)

$$A \cos(2\pi vt + \phi_2) \qquad \text{from center 2}$$

Here *A* is the amplitude of the wave scattered by each particle, and ϕ_1 and ϕ_2 are the phases of the scattered waves, relative to some arbitrary reference. The total electric field seen at *P* will then be

$$E = A \cos(2\pi vt + \phi_1) + A \cos(2\pi vt + \phi_2) \tag{9.9}$$

We may rewrite this in a more useful form by using the well-known identity: $\cos x + \cos y = 2 \cos(1/2)(x - y) \cos(1/2)(x + y)$:

$$E = \underbrace{2A \cos(1/2)(\phi_1 - \phi_2)}_{\text{amplitude}} \cos[2\pi vt + (\phi_1 + \phi_2)/2] \qquad (9.10)$$

This is an expression for the electromagnetic wave produced at P by the super-position of the two scattered waves. Its amplitude depends on the phase difference $\phi_1 - \phi_2$. If the scattered waves are 180 deg out of phase, the amplitude is zero; the waves destructively interfere. If the phase difference is zero, or some multiple of 360 deg, the waves reinforce. We can calculate the intensity of the combined wave by taking the square of the amplitude:

$$i_{1,2} = 4A^2 \cos^2\left(\frac{\Delta\phi}{2}\right) \qquad (9.11)$$

or, using another standard identity,

$$i_{1,2} = 2A^2(1 + \cos \Delta\phi)$$
$$= 2i(1 + \cos \Delta\phi) \qquad (9.12)$$

where i is the intensity scattered by one particle. Equation (9.12) states that the intensity we shall see from a pair of scatterers depends on the direction from which we observe them, ranging† from 0 to $4i$.

The situation then is as follows: For given locations of two scatterers, the intensity of scattering will be greater or less than the sum of the individual intensities, depending on the direction of observation. Now let us suppose that the two scatterers are free to move at random with respect to one another. In this case, all values of $\Delta\phi$ will be equally probable over a period of time. Then the time-averaged cosine of $\Delta\phi$ will be zero, and $i_{1,2} = 2i$. This may easily be generalized to a result for n scatterers; if they are in random motion with respect to one another, the total intensity of scattering will be the sum of the individual scattering intensities. This result obviously applies to the scattering by an ideal gas; we shall later apply it to dilute solutions. For the moment, however, let us turn to the behavior of condensed phases.

Consider, as in Figure 9.4, the scattering from a perfect crystal. It is easy to show, from the above arguments, that the scattering of long-wavelength radiation (long, as compared to the lattice spacing) will be zero, except in the forward direction. Consider any one scatterer i, observed from a given point, P. If the crystal is large enough and the lattice spacing small enough, there will always be another scattering center at some point j, which is so located that the phases of the scattered waves differ by some odd multiple of 180 deg. In other

---

† The intensity averaged over all directions is, of course, $2i$, corresponding to the fact that the energy lost to the incident beam is twice that for a single scatterer.

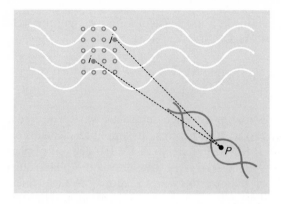

Figure 9.4

Scattering from a perfect crystal. If the crystal is large enough and the interparticle spacing small enough, there will be, for every scatterer, i, another j that is so placed that at P their scattered waves exactly cancel.

words, the crystal can be considered as made up of scattering elements that cancel one another. The only exception to this rule will be for scattering in the direction taken by the incident beam. Here the scattered waves will always be in phase, and reinforcement must occur. For a pure liquid, the situation will be similar. We can consider dividing the liquid into cells much smaller than λ (Figure 9.5). Each of these cells may be considered a scattering center. If they

Figure 9.5

Scattering from a liquid. The fluid has been divided into cells small compared to λ. Each can contain a number of molecules. Each cell is considered to be a scattering center. Suppose i and j are so placed that their scattering at point P is out of phase (as in Figure 9.4). If i and j contained equal numbers of molecules (had equal scattering power), their scattering would cancel at P. But fluctuations in density prevent complete cancellation from occurring. Thus some light is scattered.

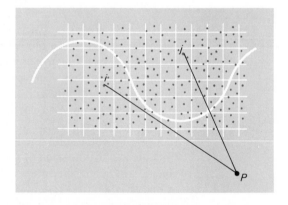

were all of equal density, and thus of equal scattering power, the argument for a crystal would apply: no scattering except in the direction of the beam. But in a real liquid, *fluctuations* in density occur from instant to instant; thus the densities in a pair of cells that are so located as to scatter out of phase will not in general be quite equal. Hence their scattering will not quite cancel. Thus, because of local fluctuations in density, a pure liquid will scatter somewhat more than a perfect crystal, though not nearly so much as the sum of the individual molecular scattering powers. Of course, once again the scattering in the forward direction will be undiminished, since phase differences are zero.

9.3   FORWARD SCATTERING AND THE INDEX OF REFRACTION†

What will be the effect of these forward-scattered waves? Consider the forward scattering from a thin layer of scatterers (a slice of a crystal or a thin liquid layer, for example), as shown in Figure 9.6. By a rather involved argument (see W. H. Kauzmann, 1957), it can be shown that the scattered wavelets combine, in the forward direction, to yield a wave 90 *deg out of phase* with the incident wave.

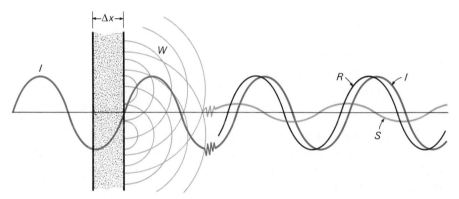

Figure 9.6   The relationship between forward scattering and refraction. The incident wave, *I*, after passing through a thin slab of matter, is accompanied by a forward scattered wave, *S*, which is 90° out of phase with it. This forward-scattered wave can be considered the composite of all the wavelets, *W*, scattered by the particles in the slab. The resultant, *R*, of the incident and forward-scattering waves is the observed emergent wave. It is retarded a little in phase.

The effect of adding this to the incident wave is shown in Figure 9.6; it is as if the radiation has been *retarded* in passing through the thin slab of matter. To describe this quantitatively, we argue as follows: The emerging wave will be the sum of a cosine term (the incident wave) and a sine term (the forward scattered wave, 90 deg out of phase). The amplitude of this second component should depend on the thickness, Δx, of the slab of matter and the number, N, of scatterers per cubic centimeter contained in it: The electric field of the emerging radiation can be written as

$$E = E_0[\cos 2\pi v(t - x/c) - \Gamma \, \Delta x \sin 2\pi v(t - x/c)] \qquad (9.13)$$
$$\text{original wave} \qquad \text{forward-scattered wave}$$

where the proportionality constant Γ turns out to be

$$\Gamma = \frac{4\pi^2 N\alpha}{\lambda} \qquad (9.14)$$

† This section is to some extent a digression. The reader who is interested solely in the use of light scattering to study macromolecules should proceed to Section 9.4.

The result can be depicted as a wave *retarded in phase* by an amount depending on $\Gamma \Delta x$. It is as if the light beam required longer to pass through the slab than it would have taken to pass through an equal distance in vacuum. This is the origin of the "lower velocity" of light in material bodies. Algebraically, we can handle this by defining a retardation time, τ, by

$$\Gamma \Delta x = \tan(2\pi\nu\tau) \tag{9.15}$$

Therefore,

$$E = E_0[\cos 2\pi\nu(t - x/c) - \tan 2\pi\nu\tau \sin 2\pi\nu(t - x/c)]$$

$$= \frac{E_0}{\cos 2\pi\nu\tau}[\cos 2\pi\nu(t - x/c) \cos 2\pi\nu\tau - \sin 2\pi\nu(t - x/c) \sin 2\pi\nu\tau] \tag{9.16}$$

$$= \frac{E_0}{\cos 2\pi\nu\tau} \cos 2\pi\nu(t + \tau - x/c) \tag{9.17}$$

Since $2\pi\nu\tau$ will be small, $\cos 2\pi\nu\tau \simeq 1$. Equation (9.17) represents the emergent wave as having been retarded by a time τ in passing through each slab of thickness Δx. If we define a new *apparent velocity* c', the extra time τ to span the layer Δx is given by

$$\tau = \frac{\Delta x}{c'} - \frac{\Delta x}{c} \tag{9.18}$$

Setting the refractive index $n = c/c'$,

$$\tau = (n - 1)\frac{\Delta x}{c} \tag{9.19}$$

But since $2\pi\nu\tau$ is small, $\tan 2\pi\nu\tau \simeq 2\pi\nu\tau$. Therefore,

$$\frac{\Gamma\Delta x}{2\pi\nu} \cong (n - 1)\frac{\Delta x}{c} \tag{9.20}$$

or

$$n - 1 = \frac{\Gamma c}{2\pi\nu} = \left(\frac{c}{2\pi\nu}\right)\left(\frac{2\pi N\alpha 2\pi\nu}{c}\right)$$

$$= 2\pi N\alpha \tag{9.21}$$

This is one form of the equation relating refractive index to molecular polarizability. A more exact treatment, which takes into account the fact that the electric field in the medium is itself altered by the dielectric properties of the medium, gives

$$\frac{n^2 - 1}{n^2 + 2} = \frac{4}{3}\pi N \alpha \tag{9.22}$$

or, to a fair approximation, since $n \simeq 1$,

$$n^2 - 1 = 4\pi N \alpha \tag{9.23}$$

9.4   RAYLEIGH SCATTERING FROM SOLUTIONS OF MACROMOLECULES

We are now in a position to consider the topic of primary interest to us, the scattering produced by a solution of macromolecules. We shall be interested in the excess scattering over that given by solvent alone; this is what will tell us about the macromolecules. There are two ways in which the problem can be approached. The more exact method involves a thermodynamic analysis of concentration fluctuations, similar to the density-fluctuation analysis suggested for the scattering by pure liquids. A simpler, and in most cases quite satisfactory, approach is to consider the solute molecules in a dilute, ideal solution to behave as independent scatterers, as do the particles in an ideal gas. Further, we shall assume for the moment that the particles are small compared with the wavelength. The scattering under these conditions (independent small particles) is called Rayleigh scattering.

Going back to Equation (9.7), we see that the scattering from each particle depends on the polarizability of the solute particle. The polarizability can, by the analysis given in Section 9.3, be related to a more accessible quantity, the excess refractive index of the solution over that of solvent. If N is the number of solute particles per cubic centimeter, Equation (9.23) yields

$$n^2 - n_0^2 = 4\pi N \alpha \tag{9.24}$$

where α is now the excess polarizability of a solute particle over the solvent it displaces. This rearranges to

$$(n - n_0)(n + n_0) = 4\pi N \alpha \tag{9.25}$$

$$\alpha = \frac{(n + n_0)}{4\pi} \frac{(n - n_0)}{C} \frac{C}{N} \tag{9.26}$$

In Equation (9.26), we have multiplied and divided by C, the concentration in grams per cubic centimeter. The quantity $(n - n_0)/C$ is the *specific refractive index increment* of the solute; for a linear variation of n with C, we may write it as dn/dC. Also, $C/N = M/\mathcal{N}$, where $\mathcal{N}$ is Avogadro's number. For dilute solutions $n + n_0 \simeq 2n_0$; therefore,

$$\alpha = \frac{n_0}{2\pi}\left(\frac{dn}{dC}\right)\frac{M}{\mathcal{N}} \tag{9.27}$$

Substituting for α in Equation (9.7), we obtain

$$\frac{i}{I_0} = \frac{2\pi^2 n_0^2 (dn/dC)^2 M^2}{\lambda^4 r^2 \mathcal{N}^2}(1 + \cos^2\theta) \tag{9.28}$$

All of this has been derived to calculate the intensity of scattering from one particle. If we have N particles per cubic centimeter, where $N = C\mathcal{N}/M$, we obtain N times as much, which we shall call i_θ:

$$\frac{i_\theta}{I_0} = \frac{2\pi^2 n_0^2 (dn/dC)^2 (1 + \cos^2\theta)}{\lambda^4 r^2 \mathcal{N}}CM \tag{9.29}$$

Equation (9.29) says that the excess scattering produced by a solution containing a weight concentration C of particles of molecular weight M depends on the product CM. It also depends on the angle with respect to the incident beam, θ, but is symmetrical as regards forward and backward scattering. We can define a quantity R_θ, the Rayleigh ratio,† which contains some of the experimental parameters:

$$R_\theta = \frac{i_\theta}{I_0}\frac{r^2}{1 + \cos^2\theta} \tag{9.30}$$

Then

$$R_\theta = \frac{2\pi^2 n_0^2 (dn/dC)^2}{\mathcal{N}\lambda^4}CM = KCM \tag{9.31}$$

where

$$K = \frac{2\pi^2 n_0^2 (dn/dC)^2}{\mathcal{N}\lambda^4} \tag{9.32}$$

These equations show that light-scattering measurements can be used for molecular-weight determination. With real solutions the equations must be modified to take into account the nonideality of the solution. Ideality was implied when we said that the scattering of N particles was N times the scattering from a single particle; this will be true only if the particles are entirely independent of one another. If there are correlations between their positions (if even, for example, the center of one particle cannot penetrate into the space occupied by another), the scattering will be modified. A precise calculation can be carried out on the basis of the thermodynamic theory of concentration fluctu-

---

† Note that R_θ is being defined so as to be independent of θ for Rayleigh scattering. This will *not* be the case for scattering from large particles.

ation [see P. Debye (1951)]. The result can be expressed simply; for an ideal solution, Equation (9.31) applies and may be written

$$\frac{KC}{R_\theta} = \frac{1}{M} \tag{9.33}$$

whereas for a nonideal solution

$$\frac{KC}{R_\theta} = \frac{1}{M} + 2BC + \cdots \tag{9.34}$$

where B is the second virial coefficient.† Obviously, scattering must generally be measured at several concentrations and extrapolated to $C = 0$.

If the solute is heterogeneous, the light-scattering method must give an average molecular weight. In fact, it is the *weight average* that is obtained. This follows from the fact that at low concentration the total R_θ must be the sum of the contributions from the various solute components, $R_{\theta i}$:

$$R_\theta = \sum_i R_{\theta i} = \sum_i KC_i M_i = K \frac{\sum_i C_i M_i}{\sum_i C_i} C = K\overline{M}_w C \tag{9.35}$$

where we have assumed K (and therefore dn/dC) to be the same for all components and have made use of the identity $\sum_i C_i = C$, where C is the total weight concentration.

Equation (9.34) indicates that to measure the molecular weight we must determine R_θ at a number of concentrations and extrapolate KC/R_θ to $C = 0$.

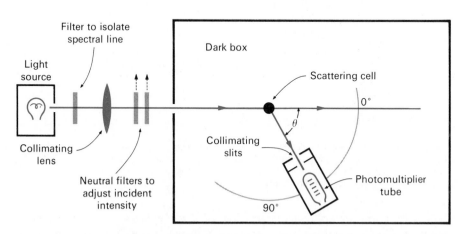

Figure 9.7 A schematic sketch of a light-scattering photometer. The photomultiplier is mounted on a turntable so that it can be rotated to observe at any angle θ. The incident beam intensity (reduced by neutral filters) is compared to the intensity scattered at various angles.

† Note the similarity of Equation (9.34) to the equation that describes the osmotic pressure of nonideal solutions.

The measurement of R_θ is usually carried out in a *photometer* similar to that diagrammed in Figure 9.7. The intensity of light scattered at a given angle (90 deg, for example) is compared with the intensity of the incident light. A complication arises from the fact that the solution must be scrupulously clean; small amounts of dust will contribute as extremely large molecules to the average in Equation (9.35), leading to serious errors. Therefore, solutions must be carefully filtered or centrifuged before use.

In addition to R_θ values, calculation of the molecular weight requires a knowledge of dn/dC. Since the difference in refractive index between a dilute solution and pure solvent is very small, a *differential* refractometer is commonly used.

9.5 SCATTERING BY LARGER PARTICLES

So far, we have assumed that the greatest dimension of the scattering particle is small compared to the wavelength of the light. If this is not true, the light-scattering experiment becomes more interesting and informative. Examine Figure 9.8. The incident light is shown inducing dipole oscillation at a pair of

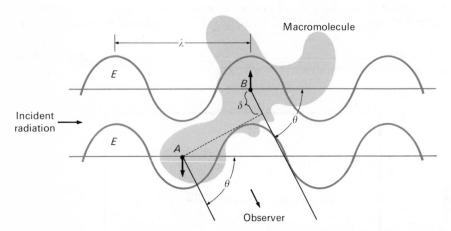

Figure 9.8   Scattering from a macromolecule that is large compared to λ. Two points from which scattering occurs are shown at A and B. The phase of the radiation (and hence of the induced dipoles) is clearly different. Also, the two points are at different distances from the observer.

points that are an appreciable fraction of a wavelength apart; the dipoles within a given molecule are oscillating out of phase. Since the scattering centers within a given large molecule are (more or less) fixed with respect to one another, we can no longer consider them to be independent scatterers. We must therefore, take into account interference between these scattering centers.

The calculation of this effect for a macromolecule must include the interference between light scattered from all pairs of scattering points within the molecule. Since we also have to consider all orientations of a molecule that is assumed to be in rapid Brownian movement the problem is complicated. The result, which is of general validity and will be encountered in a number of situations, may be stated as follows: We call $P(\theta)$ the ratio of the scattered intensity at an angle θ to that which would be observed if the particle had the same molecular weight (same number of scattering dipoles) but infinitesimal dimensions compared to λ. Then $P(\theta)$ for any aggregate of scatterers turns out to be†

$$P(\theta) = \frac{1}{N^2} \sum_{i=1}^{N} \sum_{j=1}^{N} \frac{\sin hR_{ij}}{hR_{ij}} \tag{9.36}$$

where N is the number of scattering centers within the molecules (we could take them to be the atoms), R_{ij} is the distance between any pair of centers i and j, and h is given by

$$h = \frac{4\pi}{\lambda} \sin \frac{\theta}{2} \tag{9.37}$$

where λ is the wavelength of light in the medium (that is, $\lambda = \lambda_0/n$, where n is the refractive index and λ_0 the wavelength in vacuum). Equation (9.36) is a double sum that extends over all pairs of scattering centers. To show the behavior of $P(\theta)$ in limiting cases, let us represent $(\sin x)/x$ by a Taylor's series. The familiar expansion for $\sin x$ at small x, when divided through by x, gives

$$\frac{\sin hR_{ij}}{hR_{ij}} \simeq 1 - \frac{(hR_{ij})^2}{6} + \frac{(hR_{ij})^4}{120} + \cdots \tag{9.38}$$

This immediately yields several results. As either $h \longrightarrow 0$ or $R_{ij} \longrightarrow 0$, we have

$$P(\theta) \longrightarrow \frac{1}{N^2} \sum_{i=1}^{N} \sum_{j=1}^{N} (1) = \frac{1}{N^2} N^2 = 1 \tag{9.39}$$

Thus at very low angles, at very long wavelengths, or for very small particles, Rayleigh scattering is approached. The situation at $\theta = 0$, in fact, can be inferred from Figure 9.8; as usual there is no phase difference in light scattered directly in the forward direction.

At angles greater than zero, $P(\theta)$ will always be less than unity, which means that the scattering intensity at these angles will always be less for an extended particle than for a compact particle of the same weight. A typical graph of the angular dependence of light scattering for a large particle is shown as the dotted line in Figure 9.2(c). We have multiplied the $(1 + \cos^2 \theta)/2$ factor by $P(\theta)$.

It may appear that this effect of particle size is merely an exasperating complication, but, in fact, it allows us to determine particle dimensions from light

---

† The derivation of this equation is beyond the scope of this book. See the reference to Tanford's *Physical Chemistry of Macromolecules*. It might be worthwhile to do Problem 4 at this point to gain some idea of how Equation (9.36) is derived.

scattering. Suppose that the particles are in such a size range that we need consider only the first two terms in Equation (9.38). Then

$$P(\theta) = \frac{1}{N^2} \sum_{i=1}^{N} \sum_{j=1}^{N} (1) - \frac{h^2}{6N^2} \sum_{i=1}^{N} \sum_{j=1}^{N} R_{ij}^2 \qquad (9.40)$$

The first term in Equation (9.40) is unity, and the sum in the second has already been encountered in another form in Chapter 2. The *radius of gyration* of a particle, R_G, can be expressed in terms of the distances between pairs of mass units in the molecule:

$$R_G^2 = \frac{1}{2N^2} \sum_{i=1}^{N} \sum_{j=1}^{N} R_{ij}^2$$

Therefore,

$$P(\theta) = 1 - \frac{h^2 R_G^2}{3} + \cdots = 1 - \frac{16\pi^2 R_G^2}{3\lambda^2} \sin^2 \frac{\theta}{2} + \cdots \qquad (9.41)$$

There has been no assumption made about the shape of the particle in this derivation. Thus the angular dependence of light scattering can give us unambiguously the quantity R_G. This will only be useful if the particle is fairly large, as can be seen readily from Equation (9.41). If the radius of gyration is less than one fiftieth of the wavelength (that is, less than about 80 Å for visible light in water), the deviation of $P(\theta)$ from unity will not amount to more than a small percentage at any angle. Therefore, measurement of the angular dependence of the scattering of visible light will be useful only for the larger macromolecules.

Since the scattering for a real solution of large macromolecules depends on both angle and solute concentration, it is evident that R_θ must be measured as a function of both variables. If we recall that the definition of $P(\theta)$ says

$$R_\theta(\text{real particle}) = P(\theta) \times R_\theta(\text{same particle, if it showed Rayleigh scattering})$$

then Equation (9.34) must be rewritten as

$$\frac{KC}{R_\theta} = \frac{1}{P(\theta)} \left\{ \frac{1}{M} + 2BC \right\} \qquad (9.42)$$

or, using the approximate form given in Equation (9.41),

$$\frac{KC}{R_\theta} = \frac{1}{1 - (16\pi^2 R_G^2 / 3\lambda^2) \sin^2 (\theta/2)} \left\{ \frac{1}{M} + 2BC \right\}$$

$$\simeq \left\{ 1 + \frac{16\pi^2 R_G^2}{3\lambda^2} \sin^2 \frac{\theta}{2} \right\} \left\{ \frac{1}{M} + 2BC \right\} \qquad (9.43)$$

In the case of heterogeneous samples, M is to be replaced by $\overline{M}_w$. Evidently, to obtain M it is necessary to extrapolate KC/R_θ to zero angle *and* zero concentration. In the method devised by Zimm (1948), both extrapolations are made on the same graph. Figure 9.9 shows a Zimm plot for data on the light

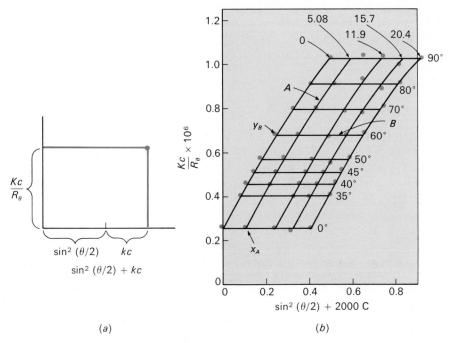

(a)   (b)

Figure 9.9   (a) How a Zimm plot is constructed. One measures the distance $\sin^2(\theta/2) + kc$ on the abcissa, KC/R_θ on the ordinate. (b) Light scattering of DNA. The numbers on top refer to concentrations; those on the right to angles. Data taken from P. Doty and B. Bunce, *J. Am. Chem. Soc.*, **74**, 5029, (1952). Copyright © 1952 by American Chemical Society.

scattering of a macromolecule. The quantity KC/R_0 has been graphed versus $(\sin^2 \theta/2 + kC)$, where k is an arbitrary constant, here taken to be 2,000 to give a convenient scale. For each measurement of R_θ (at a particular value of θ and C), we measure a distance $\sin^2(\theta/2) + kC$ on the abscissa and plot the corresponding value of R_θ as the ordinate. Now consider a series of points at a given concentration but different angles. Such a series will lie along a line such as A. The value of KC/R_0 that will be approached at *zero angle* and *this concentration* is given by the point at which the line A has the abscissa value kC—the point x_A in this case. A series of such points have been obtained, one at each concentration. Together, they lie on a line (denoted by 0°) that represents the scattering of this sample at various concentrations and zero angle. The equation of this line is, from Equation (9.43),

$$\frac{KC}{R_\theta}\bigg|_{\theta=0} = \frac{1}{M} + 2BC \qquad\qquad (9.44)$$

and thus one finds M and B from the intercept and slope.

Now suppose that instead of proceeding as above, we choose a series of points at a given scattering angle but different concentrations. If we extrapolated such a series such as the line B to the point where $C = 0$ [where $\sin^2(\theta/2) + kC = \sin^2(\theta/2)$], the point y_B would be obtained. This represents the value of KC/R_θ approached at a given θ as $C \longrightarrow 0$. It is easy to see that the series of zero-concentration points (denoted by 0) is described by the equation

$$\left.\frac{KC}{R_\theta}\right|_{\theta=0} = \frac{1}{P(\theta)}\frac{1}{M} = \frac{1}{M}\left\{1 + \frac{16\pi^2}{3}\frac{R_G^2}{\lambda^2}\sin^2\frac{\theta}{2}\right\} \tag{9.45}$$

The intercept gives a check on M, and the slope gives R_G.

In Table 9.1 some values of M and R_G determined by light scattering are given. Although R_G itself is unambiguous, its interpretation in terms of dimensions of the particles depends on particle shape. Table 9.2 lists some values of R_G in terms of particle dimensions. It should also be emphasized that the angular dependence of scattering will depend explicitly on shape if higher terms in the expansion in (9.40) have to be retained.

TABLE 9.1   REPRESENTATIVE DATA FROM LIGHT SCATTERING AND LOW-ANGLE X-RAY SCATTERING[a]

Material	M_w	$R_G(\text{Å})$
Lysozyme	14,100	15.2
β-Lactoglobulin	36,000	
	36,700	21.7
Serum albumin	70,000	29.8
Myosin	493,000	468
DNA sample	4×10^6	1170
TMV	39×10^6	924
Turnip yellow mosaic virus		104

[a] Underlined values are from low-angle X-ray scattering.

TABLE 9.2   RELATION BETWEEN R_G AND DIMENSIONS FOR PARTICLES OF VARIOUS SHAPES

Shape	R_G	Where:
Sphere	$\sqrt{3/5}R$	R = radius of sphere
Prolate ellipsoid	$(\sqrt{2 + \gamma^{-2}/5a})$	Axes are $2a$, $2a$, $\gamma 2a$
Very long rod	$L/\sqrt{12}$	L = lenth of rod
Random coil	$(\overline{h^2})^{1/2}/\sqrt{6}$	$\overline{h^2}$ = mean-square end-to-end distance

In summary, light scattering provides a powerful tool for the determination of particle weight and dimensions. Its use is limited at one extreme by the low scattering power of small molecules and at the other by the difficulties encountered in interpretation for very large macromolecules. If the particles are very large, Equation (9.40) no longer adequately describes the angular dependence, which then reflects the particle shape as well. If the shape is not known, or if experimental difficulties prevent measurement to θ low enough for Equation (9.40) to be valid, a clear extrapolation to $\theta = 0$ may not be possible. In practice, it is very difficult, because of diffraction effects, to measure scattering at angles less than about 5 deg.

9.6 LOW-ANGLE X-RAY SCATTERING

While the *weights* of small molecules can be determined from the scattering of visible light, the method fails if we seek to use the angular dependence to measure the *dimensions* of particles with R_G less than about 100 Å. Since the *resolving* power of the method depends on the ratio $(R_G/\lambda)^2$, an obvious solution would be to use light of shorter wavelengths. But here nature has been un-cooperative, for those materials that we would most like to investigate (proteins and nucleic acids, for example) begin to absorb very strongly a little below 300 nm. Below about 200 nm, almost everything, including water and air, absorbs very strongly, leaving a considerable region of the spectrum at the present time inaccessible to us. It is not until we reach the X-ray region that materials again become generally transparent. But here the scattering situation is very different; the interference effects, which were second order in the visible, now dominate the scattering. For example, the wavelength of the Cu-α radiation (a commonly used X radiation) is 1.54 Å, so that the distance across a ribonuclease molecule (molecular weight of 13,683) is of the order of 10λ. Clearly, an approximation such as Equation (9.40) will fail even at quite small angles, and the complete Equation (9.36) must be used. The situation is best illustrated by a specific example. Evaluation of the quantity $\sum \sum \sin hR_{ij}/hR_{ij}$ for atoms distributed uniformly in a solid sphere of radius R leads to the $P(\theta)$ function shown in Figure 9.10. For this example, we have chosen $\lambda = 1.54$ Å and $R = 15.4$ Å. The scattering is almost entirely confined to a very narrow

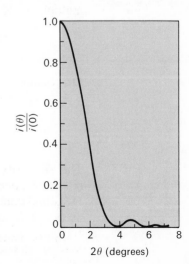

Figure 9.10

A graph of the logarithm of scattering intensity versus angle for the scattering of 1.54-Å X rays from spheres of radius 15.4 Å.

angular range. The effect of the interference of waves is emphasized by the existence of maxima and minima in the $P(\theta)$ function.

Fortunately, it is possible to collimate an X-ray beam much better than a light beam, so measurements can often be made to angles low enough to allow a decent extrapolation to $\theta = 0$. If this is done, the molecular weight can be calculated from the zero angle scattering, much as in light-scattering experiments. The theory looks a little different because it is most convenient to consider the individual electrons in the molecule as scatterers. By classical theory the intensity of X-ray scattering from a single electron is

$$i_e(\theta) = \frac{I_0}{r^2}\left(\frac{e^2}{mc^2}\right)^2\left(\frac{1 + \cos^2\theta}{2}\right) \tag{9.46}$$

where e is the electron charge, m the electron mass, and c the velocity of light. For a molecule containing n electrons, we shall have an intensity n^2 times as great, for the scattering is coherent and *amplitudes* must be added:

$$i_m(0) = \frac{I_0}{r^2}\left(\frac{e^2}{mc^2}\right)^2 n^2 \tag{9.47}$$

where θ has been taken† as 0 deg. To calculate the scattering from 1 cm³ of a solution containing N molecules per cubic centimeter, we add the *intensities* (since the molecules are independent of one another):

$$i(0) = \frac{I_0}{r^2}\left(\frac{e^2}{mc^2}\right)^2 n^2 N \tag{9.48}$$

We are interested in solutions, so we replace n by $n - n_0$, the excess electrons in a solute molecule over the volume of solvent it displaces. Also we divide and multiply by M^2:

$$i(0) = \frac{I_0}{r^2}\left(\frac{e^2}{mc^2}\right)^2 NM^2\left(\frac{n - n_0}{M}\right)^2 \tag{9.49}$$

Or, recalling that $N = \mathcal{N}C/M$, where C is the weight concentration and $\mathcal{N}$ is Avogadro's number,

$$i(0) = \frac{I_0}{r^2}\left(\frac{e^2}{mc^2}\right)^2 \mathcal{N}\left(\frac{n - n_0}{M}\right)^2 MC \tag{9.50}$$

The similarity to the light-scattering equation is obvious. The quantity $(n - n_0/M)$ depends on the chemical composition, since it is essentially a ratio of atomic numbers to atomic weights. As in the case of light scattering, it is necessary to extrapolate data to $C = 0$.

† By concentrating on the zero angle scattering here, we eliminate the problems of phase differences between scattering points.

The angular dependence of the scattering in the very low-angle range will give the radius of gyration. The extrapolation might be made by graphing $i(\theta)$ versus $\sin^2(\theta/2)$ as in light-scattering experiments, but it has become customary to attempt to extend the range of a linear graph by a slightly different method. The first two terms of Equation (9.41) are identical with those in an expansion of $e^{-h^2 R_G^2/3}$; that is,

$$e^{-h^2 R_G^2/3} = 1 - \frac{h^2 R_G^2}{3} + \frac{h^4 R_G^4}{18} - \cdots \tag{9.51}$$

For spherical particles, even the third term is identical to the expansion of the exponential. Thus to a good approximation for any particle in the low-angle range, we can write

$$P(\theta) \cong e^{-h^2 R_G^2/3} \tag{9.52}$$

or

$$\ln P(\theta) = -h^2 R_G^2/3 = -\frac{16\pi^2 R_G^2 \sin^2 \theta/2}{3\lambda^2} \tag{9.53}$$

which means that graphing $\ln i(\theta)$ versus $\sin^2(\theta/2)$ will give a straight line with intercept $\ln i(0)$ and slope $-16\pi^2 R_G^2/3\lambda^2$. In Table 9.1 some molecular weights and particle dimensions obtained by low-angle X-ray scattering are given. Even the smallest protein molecules can be investigated by this method.

The angular dependence of the scattering at somewhat larger angles becomes dependent on the shape of the particles. While the analysis has not been carried as yet to the point where all of the potentially available information can be obtained, it is possible to determine such details as the cross-sectional area and mass per unit length of elongated particles.

PROBLEMS

*1. At one time, it was conventional to describe light scattering in terms of the *turbidity*, τ. This is defined in analogy to the extinction coefficient in spectroscopy; if a beam of incident intensity I_0 passes through 1 cm^3 of solution and emerges with a smaller intensity I (the loss being by scattering), we define

$$\tau = -\ln I/I_0$$

Show that for Rayleigh scattering, $\tau = (16\pi/3)R_\theta$. (*Hint:* Integration of the scattered intensity over a sphere at distance r from the solution should give the total loss of energy, but this should also be equal to $I - I_0$ as defined above. Why?)

2. A complication in the absorption spectroscopy of very large molecules arises from the light lost by scattering. Calculate the absorbance due to *scattering* alone, for a solution of a hemocyanin, given the following data:
(a) Cell thickness in direction of beam $= 1 \text{ cm}$.
(b) Protein concentration $= 0.01 \text{ g/ml}$, $M = 3,800,000$.
(c) $n_0 = 1.33$, $dn/dC = 0.198 \text{ ml/g}$, $\lambda = 2,800 \text{ Å}$.
You may assume Rayleigh scattering. See Problem 1.

3. To visualize the effect of dust contamination on light-scattering experiments, perform the following calculation. A solution, containing 5 mg/ml of protein of $M = 100,000$ is contaminated to the extent of 0.1 percent of the protein weight by dust. The dust particles are 0.1μ in radius, with density $= 2.00$. By what percent will the 90-deg scattering be changed by this contamination, assuming Rayleigh scattering for all particles?

*4. To provide a specific example to illustrate the general problem of large particle scattering, let us consider the scattering from a dumbbell-shaped molecule containing two scattering centers separated by a distance R. To simplify the geometry, we shall require that the molecule remain in the plane of the paper but assume that it is in rapid

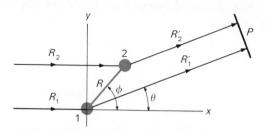

Brownian rotation in that plane. The molecule assumes an angle ϕ to the x axis, with one of its scattering centers at the center of the coordinate system. We consider incident light parallel to the x axis, scattered at an angle θ to this axis. Calculate the function $P(\theta)$ for this model, making any necessary approximations. (*Hint:* The scattering angle θ can vary from zero to π, and we must consider all orientations of the particle with respect to the x axis. The path difference at P for R'_2 and R'_1 is best expressed in terms of an angle ψ, such that $\phi = \psi + \theta/2$.)

5. The radius of gyration for a circle is just the radius of the circle. The expression for the R_G of a rod is given in the text. Suppose a DNA molecule could exist either as a rigid rod 1,000 Å long or as a circle with this circumference. Calculate the ratio of scattering at $\theta = 90$ deg for the rod form to that for the circle form. Assume that $\lambda_0 = 5460$ Å. If a 1 percent difference is detectable, would the light-scattering measurement serve to detect the breaking of circles?

6. The following data have been given [P. Outer, C. T. Carr, and B. H. Zimm, *J. Chem. Phys.*, **18**, 830 (1950)] for the light scattering of polystyrene in butanone at 20°C. The light was polarized perpendicular to the plane of measurement, that is, along the z axis. The following data are given for various angles and concentrations:

Values of $CI_0/r^2 i_\theta$

	C (g/ml)			
θ (deg)	0.312×10^{-3}	0.624×10^{-3}	1.25×10^{-3}	2.50×10^{-3}
26	1.58	1.65	1.86	2.21
36.9	1.66	1.76	1.98	2.33
66.4	1.82	1.90	2.07	2.52
90	1.95	2.08	2.24	2.69
113.6	2.17	2.24	2.47	2.88

$n_0 = 1.378$, $dn/dC = 0.214$ ml/g, and $\lambda = 5,460$ Å. Obtain R_G and $\overline{M}_w$. Be sure to use correct equations for polarized light. $P(\theta)$ does *not* depend on polarization.

7. Ribosomes (50 S) from *E. coli* have been studied by low-angle X-ray scattering. The following data were obtained, using radiation of $\lambda = 1.54$ Å:

log I (units arbitrary)	θ (rad) $\times 10^3$
1.76	1.41
1.73	2.00
1.70	2.54
1.66	3.00

Calculate the radius of gyration of 50-S ribosomes. If they are spheres, what are their diameters?

REFERENCES

Light Scattering: Fundamental Principles

Debye, P.: *J. Phys. Chem.*, **51**, 18 (1951). Analysis of scattering by fluctuation theory; a bit specialized.

Feynman, R. P., R. B. Leighton, and M. Sands: *The Feynman Lectures on Physics*, Vol. I, Addison-Wesley Publishing Company, Inc., Reading, Mass., 1963, Chap. 28–32. A remarkably lucid discussion of the fundamentals; the whole set of books is highly recommended.

Kauzmann, W. H.: *Quantum Chemistry*, Academic Press, Inc., New York, 1957, Chap. 15. Good treatment of fundamentals.

Tanford, C.: *Physical Chemistry of Macromolecules*, John Wiley & Sons, Inc., New York, 1961, Chap. 5. A more detailed exposition of the light scattering by macromolecules.

Light Scattering: Applications

Stacey, K. A.: *Light Scattering in Physical Chemistry*, Academic Press, Inc., New York, 1956. A thorough discussion, best for experimental technique.

Stacey, K. A.: in *Analytical Methods of Protein Chemistry* (P. Alexander and R. J. Block, eds.), Pergamon Press, Inc., New York, 1961. A brief review, much like the preceding in treatment.

Zimm, B. H.: *J. Chem. Phys.*, **16**, 1093, 1099 (1948). Development of Zimm's methods for analysis of angular dependence.

Low-Angle X-Ray Scattering

Beeman, W. W., P. Kaesberg, J. W. Anderegg, and M. B. Webb: *Handbuch der Physik* (S. Flugge, ed.), Vol. XXXII, Springer-Verlag, Berlin, 1957, p. 321.

TEN | CIRCULAR DICHROISM AND OPTICAL ROTATORY DISPERSION

One feature shared by a wide variety of biopolymers is the existence of helical secondary structure. The α helices of polypeptides and proteins, the single, double, and triple helices of polynucleotides, and the helical conformations of some polysaccharides are both unique in each class of compounds and important to their biological function. The detection of such structure in biopolymers and quantitative measurement of the amount and changes in amount of such structures play an important role in our understanding of the way in which these molecules function.

Unfortuately, none of the physical properties we have discussed so far are very useful for the detection of helical structure. It is true that absorption spectra are often sensitive to regular secondary structure, via hypochromism and exciton splitting, but it would be difficult to rest the case for helical conformations on this kind of evidence alone. X-ray diffraction (Chapter 11) can yield detailed information on structure but is applicable only to the solid state. What is needed is a physical property that depends explicitly on the asymmetry of structure of dissolved molecules. This is provided by the response of molecules to polarized light.

10.1   POLARIZATION OF RADIATION

An electromagnetic wave is characterized by the amplitude and orientation of its component electric and magnetic fields. As shown in Figure 10.1, the electric and magnetic field vectors will always be perpendicular to one another and perpendicular to the direction of propagation. It is by their orientation with respect to an external frame of reference (the apparatus, a molecule) that we describe the polarization. The simplest case is that of plane polarization (Figure 10.1); here the direction of propagation is along the x axis, the electric vectors point in the z direction, and the magnetic vectors point in the y direction.

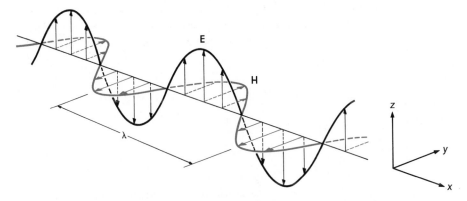

Figure 10.1   Plane-polarized radiation. See the text.

At a given point on the x axis, the field vectors oscillate with a frequency v (reciprocal seconds).

Plane-polarized light is so easy to produce and work with that the significance of other kinds of polarization is often neglected. Equally important is circularly polarized light. In Figure 10.2 we show the electric field only. At any point

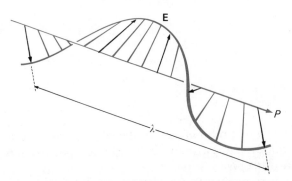

Figure 10.2   Circularly polarized radiation. Only the electric field is shown. By the convention used, this is right circular polarization, since an observer at point P would see the field rotating in a clockwise direction as the waves passed.

on the x axis the electric vector can be visualized as rotating in a plane perpendicular to that axis, making ν rotations/sec. The retardation of successive vectors along the x axis is such that the wavelength λ corresponds to one complete rotation (note in Figure 10.2). The motion of a point of a field vector is a helical path, with a velocity of propagation in the x direction of $c = \lambda\nu$. Circular polarization can be either right- or left-handed, depending on the sense of rotation with respect to the direction of propagation.

By convention, we consider light to be right circularly polarized when the electric vector, as seen looking toward the source, rotates in a clockwise direction.

If we combine left and right circularly polarized waves of equal amplitude, the resultant of the vector additions of the fields will be plane-polarized light (see Figure 10.3). This means that a plane-polarized beam can be considered the sum of two circularly polarized components. We shall use these facts later.

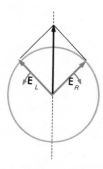

Figure 10.3

Plane-polarized light as the sum of two equal circularly polarized components. We are looking directly toward the source. The vector sum of the two field components will oscillate along the dotted line as $\mathbf{E}_L$ and $\mathbf{E}_R$ rotate.

If two circularly polarized components of unequal amplitude are combined (Figure 10.4), the result is *elliptically* polarized light. This will be characterized

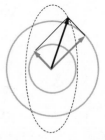

Figure 10.4
Elliptically polarized light. See the text.

by both an ellipticity (the ratio of axes of the ellipse) and the direction of the major axis. Obviously, in the extreme limits, elliptical polarization approaches either circular or plane polarization.

It is equally easy to represent a circularly polarized wave as the sum of two plane-polarized waves, 90 deg out of phase and with their electric vectors perpendicular. See Figure 10.5.

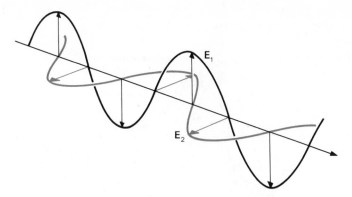

Figure 10.5   The same circularly polarized wave shown in Figure 10.2 is here represented as the sum of two plane-polarized waves. Only the electric fields are shown. Note how the electric vector, which will be the sum of E_1 and E_2, describes the same helical path as in Figure 10.2.

Finally, *unpolarized* light can be considered as a wave in which there is no fixed direction of polarization. An incoherent source, which produces a melange of waves of different polarization properties, will yield such radiation. Various optical devices may be used to separate from an unpolarized beam radiation with the several kinds of polarization characteristics described above. We turn now to the interaction of polarized light with molecules.

10.2   CIRCULAR DICHROISM AND OPTICAL ROTATION

Some, but not all substances exhibit what is called *optical activity*. This is manifested in two seemingly different but closely related phenomena: circular dichroism and optical rotation.

Circular Dichroism

If we pass circularly polarized light through a solution of an optically active substance, we shall find that the absorptivity depends on the handedness of the light. As shown in Figure 10.6, some absorption bands may absorb more strongly the left circularly polarized beam; others may absorb more strongly the right. We define the circular dichroism at a given wavelength λ as $\Delta\varepsilon = \varepsilon_L - \varepsilon_R$, the difference in extinction coefficients. A particular absorption band can be characterized by its *rotational strength*, which is essentially an integrated extinction coefficient difference over the band:

$$R = \frac{(2.303)(3000)hc}{32\pi^3 \mathcal{N}} \int \frac{\Delta\varepsilon}{\lambda} d\lambda \tag{10.1}$$

where h is Planck's constant and c the velocity of light. It is important to note

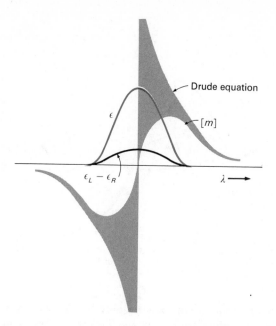

Figure 10.6
An optically active absorption band, with extinction coefficient ε, positive circular dichroism $(\varepsilon_L - \varepsilon_R)$, and an accompanying Cotton effect $[m]$. See the text. The rotational strength would be proportional to the area under the circular dichroism curve.

that circular dichroism is observed only in the wavelength regions where the substance absorbs light.

Optical Rotation

If plane-polarized radiation of *any* wavelength is passed through an optically active substance, the plane of polarization will be rotated (see Figure 10.7).

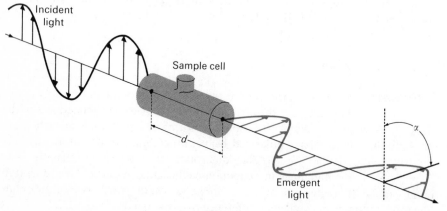

Figure 10.7   The phenomenon of optical rotation. The incident beam is plane-polarized, with electric vector in the z direction. The emergent light is still plane-polarized, but the plane has been rotated through an angle α. The solution is dextrorotatory.

The angle of rotation (α) is found to depend on the nature of the substance, the thickness of the sample (d), and the concentration (C) of the optically active substances in the sample. Thus

$$\alpha = [\alpha]\, dC \tag{10.2}$$

Here C is measured in grams per milliliter and d (by old convention) in decimeters.

The quantity $[\alpha]$ so defined is called the *specific rotation*. Ideally, it depends only on the nature of the optically active substance. Note that rotation of the plane of polarization can be either clockwise or anticlockwise as one looks toward the source of radiation. The first is termed dextrorotation and is given a positive sign; the second is termed levorotation and is given a negative sign.

The relationship between circular dichroism and optical rotation is analogous to the relationship between the absorption and refraction of unpolarized light. It will be recalled, from Chapters 8 and 9, that the interaction of an electromagnetic wave with the electrons in a chromophore can be of two kinds: Far from an absorption band (resonance frequency) there are only induced oscillations, with accompanying scattering, and the forward scattering gives rise to the effect of refraction. Near a resonance frequency (absorption band) there is as well a loss of energy to the molecule; in the terminology of quantum mechanics, we say that there is a finite probability of excitation to a higher energy state of the molecule. Now it should be clear that circular dichroism is an absorptive phenomenon. It remains to show that optical rotation is the consequence of refractive effects. If an optically active substance shows a difference in absorption between left and right circularly polarized light, it should be expected that there will be, at all wavelengths, a difference in refractive index for the two components. At the beginning of this chapter we saw that plane-polarized light could be represented as the sum of two circularly polarized beams, of equal amplitude and opposite sense (Figure 10.3). Consider what happens when a plane-polarized ray passes through an optically active substance: The left

Figure 10.8

Optical rotation as the consequence of a different refractive index for left and right circularly polarized components. The situation is the same as in Figure 10.7. The substance has a higher refractive index for L than R; therefore the L component is retarded more and has been set back with respect to R. The sum of the components is still plane-polarized, but the plane has been rotated.

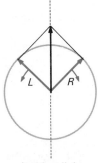
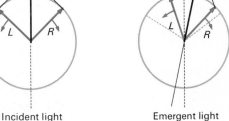

Incident light   Emergent light

and right circularly polarized components will each be retarded by refraction, but one will be retarded more than the other. The effect is shown, in Figure 10.8, to be a rotation of the plane of polarization. Thus we may state the relationship between circular dichroism and optical rotation succinctly: Circular dichroism results from a difference in absorption of L and R. Optical rotation results from a difference in refractive index for L and R.†

The analogy of optical rotation to refractive index is made stronger when we consider the wavelength dependence, the optical rotatory dispersion.‡ Any refractive effect is a result of induced dipole oscillation, and a driven oscillator behaves in a very characteristic way as the frequency is varied. At very low driving frequencies (long wavelengths) the oscillator follows in phase with the driving force. As resonance is approached, the response of the oscillator becomes stronger; if the oscillator is not damped, the response becomes infinite at resonance. In any event it becomes very large, and the original phase relation is lost. At very high frequencies, the oscillator lags 180 deg out of phase with the driving field. The effect of this phase shift is a shift of the phase of the forward scattered radiation by 180 deg. In refraction, the consequence is this: The transmitted beam now appears to lead, rather than follow behind, the incident beam, and the velocity of light appears to be larger in the medium than in vacuo. In optical rotation the effect is a change of sign of the rotation. If the rotation was positive at long wavelength, it passes through zero and becomes negative at short λ. Such a change in sign in the neighborhood of an absorption band is called a Cotton effect (Figure 10.6). For each optically active band, there will appear a circular dichroism band and a Cotton effect in the optical rotatory dispersion.

In describing optical rotatory dispersion curves, the quantity most frequently used is not the specific rotation $[\alpha]$ but the molar rotation $[m']$. This is defined as

$$[m'] = [\alpha]\frac{3}{n^2 + 2}\frac{M}{100} \tag{10.3}$$

where n is the refractive index of the medium and M the solute molecular weight. This definition has two advantages; it puts rotation on a molar (rather than weight) basis, and the factor $(3/n^2 + 2)$ corrects for a minor effect of the polarizability of the medium on the effective field seen by the molecules. In

---

† Note in Figure 10.8 that if there had been a difference in absorption of L and R as well (that is, if the wavelength corresponded to a circular dichroism band), the emergent component vectors would have been of unequal amplitude. The result would have been not only rotation of the plane of polarization, but also conversion from plane to elliptically polarized light. The molar ellipticity $[\theta]$ may be used as a measure of circular dichroism. In fact, a simple relationship exists:

$$[\theta] = 3,300\Delta\varepsilon$$

One finds experimental values expressed either in terms of $[\theta]$ or $\Delta\varepsilon$. If the dimensions of $\Delta\varepsilon$ are liter/cm · mole, $[\theta]$ is in deg cm²/decimole.

‡ "Optical rotatory dispersion" is a long phrase. Often one finds this abbreviated as ORD; similarly, circular dichroism is often referred to as CD.

working with biopolymers, the M taken is often the average residue weight; the quantity $[m']$ is then called the mean residue rotation.

The variation of $[m']$ with wavelength, at regions not too close to an absorption band, can be described by the Drude equation:

$$[m'] = \frac{96\pi \mathcal{N}}{hc} \frac{R\lambda_0^2}{\lambda^2 - \lambda_0^2} \tag{10.4}$$

where h is Planck's constant, c the velocity of light, R the rotational strength, and λ_0 the wavelength of the band. This kind of wavelength dependence is typical of the response of an undamped oscillator driven by a periodic field. As can be seen in Figure 10.6, the Drude equation becomes a poor approximation near λ_0, since it predicts infinite response. Inclusion of a *damping term* would yield a curve like the observed Cotton effect curve.

If a number of transitions of wavelength λ_{k0} and rotational strengths R_k are contributing to the optical rotation at a given wavelength, we can simply sum their contributions:

$$[m'] = \frac{96\pi \mathcal{N}}{hc} \sum_k \frac{R_k \lambda_{k0}^2}{\lambda^2 - \lambda_{k0}^2} \tag{10.5}$$

This expression will be accurate only at wavelengths far from *any* of the λ_{k0}.

Since the phenomena of circular dichroism and optical rotatory dispersion are just different manifestations of the same underlying behavior, it is to be expected that one could be predicted from the other. This is, in fact, true, and there exist equations (the Kronig–Kramer relations) that allow us to pass from CD data to ORD data and vice versa. For example, the equation that allows one to go from the CD spectrum to the ORD value† at a particular λ is

$$[m']_\lambda = 2.303 \frac{9{,}000}{\pi^2} \int_0^\infty \Delta\varepsilon_{\lambda'} \frac{\lambda'}{\lambda^2 - (\lambda')^2} \, d\lambda' \tag{10.6}$$

It may be wondered, since the two phenomena of CD and ORD are so closely related, why techniques have been developed for measuring both. Note, however, that each has its advantages. Circular dichroism curves exhibit higher resolution since each band corresponds to a particular absorption band. On the other hand, the ORD curve corresponding to each transition is spread over a very large region of the spectrum (see Figure 10.9). This means that resolution will be very difficult if a complex series of optically active bands is involved. However, there is an advantage with ORD measurements when the optically active transitions lie at wavelengths inaccessible to the instruments available.

† The reader may wonder why it is necessary to measure both CD and ORD, given the existence of relations such as Equation (10.6). Note, however, that to calculate exactly the optical rotation at any one wavelength, the entire CD spectrum must be known, since the integral in (10.6) extends from $\lambda' = 0$ to $\lambda' = \infty$. In practice, approximate results can often be obtained by considering only the nearest strong bands.

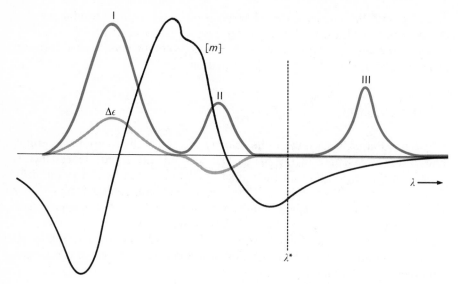

Figure 10.9   A sketch of the absorption spectrum, CD spectrum, and ORD spectrum for a hypothetical substance. Several features should be noted. (1) Not all transitions need show measurable optical activity in a complex molecule. Here band III is virtually inactive. (2) The CD spectrum readily separates the positive dichroism of band I from the negative dichroism of band II. (3) Although band II exhibits a negative Cotton effect and band I a positive Cotton effect, the ORD curves overlap so much as to make interpretation difficult. (4) If experimental difficulties preclude measurements below λ^*, ORD would give some information but CD none.

Since the ORD curve extends far beyond the absorption band, it is possible to study, for example, the rotatory properties of far-ultraviolet transitions by examining the ORD in the near ultraviolet or even in the visible region of the spectrum.

10.3 MOLECULAR BASIS OF ROTATORY POWER

Why do some substances behave in this way? Why do they exhibit different absorption and refractivity of left and right circular polarization? A clue can be found by considering the two molecules shown in Figure 10.10. These substances are very similar in structure. Both have absorption bands in the near ultraviolet, involving electron displacements in the conjugated ring systems. Yet coronene is optically inactive, whereas hexahelicene exhibits very large circular dichroism and optical rotation. Why is this?

One important and obvious difference between these molecules is that coronene is planar and has a plane of symmetry, whereas hexahelicene is asymmetric—in fact, it is helical. Let us consider what happens when an electron is driven in a helical path. In Figure 10.11 is shown an idealized molecule, which could be thought of as a prototype for any right-handed helical structure.

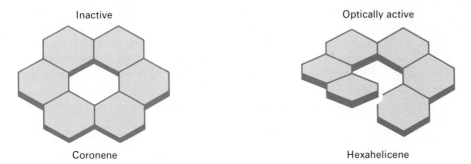

Inactive

Optically active

Coronene

Hexahelicene

Figure 10.10 Two similar substances, one of which is optically active, and the other inactive. See the text.

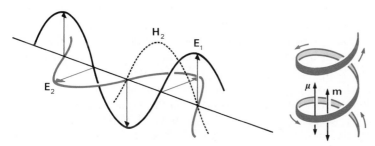

Figure 10.11 The interaction of a circularly polarized wave with a helical molecule. Electron paths are shown by arrows on the helix. The induced electric and magnetic moments parallel to the helix axis are denoted by μ and m. The magnetic field associated with the leading component of the radiation ($\mathbf{H_2}$) is indicated by a dotted line.

The helical molecule is pictured as interacting with right circularly polarized light. For the purpose of this discussion, we have chosen to represent circularly polarized light as the sum of two perpendicular plane-polarized components, differing in phase by 90 deg. Now consider the possible interaction of the radiation with the molecule. In the first place, the electric field of the radiation will induce an oscillating dipole on the helical path. Note that since the path is a helix, even an electric field parallel to the helix axis can induce such motion. (The electron goes up and down, as well as going around—there is a component of the induced electric dipole parallel to the helix axis). But still another important consequence arises from the helical structure. An oscillating magnetic field parallel to the helix axis will induce a current in the helix. This will depend, according to the laws of electromagnetism, on the derivative of the magnetic field. The magnetic field of circularly polarized light has a component parallel to the electric field (see the dotted curve in Figure 10.11). This is 90 deg out of phase with the electric field parallel to the helix axis. But its derivative will be *in* phase. Thus both the electric and magnetic fields of the radiation will contribute to an electron displacement on a helical path. In other words, electrons circulating about a helix create both an electric dipole moment and a magnetic dipole moment in the axis direction. But (and here is the crux of the matter) the phase relationship of the two plane-polarized components will differ by 180 deg for left and right circularly polarized light (at this point, see which phase the component E_1 would have to have in Figure 10.11 to yield *left* circularly polarized light). This means that whereas for one sense of polarization the electric and magnetic fields will be acting in concert on the electron, in the other sense of circular polarization they will be opposed. Thus we can qualitatively show that a molecule such as that in Figure 10.11 should exhibit different absorptivities and different refractive indices for left and right circularly polarized light. Of course, the argument is sketchy and incomplete. We should for example, consider different orientations of the molecule. But note that even when turned end for end, a right helix is still a right helix.

Another critical point emerges. If we had considered a helix of the opposite sense, the only difference would be that electron circulation would proceed in the opposite direction for a given electric field. This would reverse the magnetic dipole moment; therefore, a left-hand helix ought to display optical rotatory effects just the opposite to those given by a right-hand helix.†

The above argument is qualitative. For a more detailed treatment, the reader should consult W. Kauzmann (1957). But the result of a quantum-mechanical treatment can be stated very compactly. We have already encountered the concept of a transition dipole moment (μ), which we may associate with the charge displacement in a transition. If charge rotation occurs as well, an analogous quantity, the magnetic transition moment (m), can be defined. The existence of optical rotatory power, from a quantum-mechanical point of view, depends on the magnitudes of these moments and their relative orientation. In fact, the rotational strength is given by

$$R = \mathrm{Im}(\mu \cdot m) \tag{10.7}$$

where Im means the "imaginary part of" the dot product of the two vectors. (This nomenclature is necessary because the moments are complex vectors and their phase differs by 90 deg.) For purposes of calculation, we may say in most instances that R is given by $m\mu \cos \theta$, where m and μ are the vector magnitudes and θ is the angle between them. Thus, if the electric and magnetic moments have components in parallel, R is positive. If they are opposed, R is negative. If they are perpendicular, $R = 0$. This takes care of coronene very nicely, for even if the electron motion in such a planar molecule is a curve in the plane, the electric moment will lie in the plane and the magnetic moment will be perpendicular to it (Figure 10.12). In a simple physical sense the electric and magnetic fields are neither able to work together nor opposed upon the electron. Thus coronene is optically inactive.

† A molecule like hexahelicene can exist in either right- or left-hand helical forms, A *racemic* mixture, containing equal amounts of the two forms, would exhibit no optical activity.

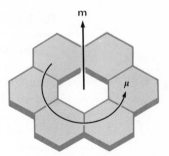

Figure 10.12

A molecule such as coronene, with a plane of symmetry, is optically inactive, since the magnetic and electric moments must be perpendicular.

Finally, it should be emphasized that it is a plane of symmetry that is critical for the absence of optical rotation. For even if substituents were placed above and below the plane of the coronene ring, with the ring plane as a mirror plane, and even if out-of-plane helical electron motions occurred in such substituents, these motions would be of opposite sense and thus cancel insofar as rotatory power was concerned. The *asymmetric carbon atom*, mentioned so often by carbohydrate chemists, now appears as a special case. If four different substituents are attached to a carbon, the static electric field they generate must be asymmetric. Then the path of any displaced electron must be a portion of some kind of a helix; therefore optical activity should result.

10.4 ROTATORY BEHAVIOR OF MACROMOLECULES

It should be clear by now that optical activity (optical rotation, circular dichroism) is observed when and only when the environment in which a transition occurs is asymmetric. In macromolecules there are at least three kinds of asymmetry that can lead to optical activity:

1. The primary structure may be inherently asymmetric. For example, transitions involving electrons in the vicinity of the α carbons of amino acids and polypeptides will be optically active because of the inherent asymmetry at this point; the α carbons of most amino acids have four different substituents.
2. The secondary structures of many biopolymers are helical. This may result in optical activity for electronic transitions either in the chain backbone or in helically arrayed side groups.
3. The tertiary structure of a macromolecule may be such that an inherently symmetric group is thrust into an asymmetric environment. For example, transitions involving the ring electrons in tyrosine are normally only weakly optically active, but in some globular proteins the surroundings of a buried tyrosine are such as to produce asymmetrical electrical fields about the ring. These may distort the electron displacement in the transition so as to lead to strong optical activity.

It is the second example above, the influence of helical secondary structure on optical activity, that has received the greatest attention to date.

The behavior of helical structures has already been alluded to above. The electron-on-a-helix analog is useful in describing the behavior of some kinds of helical polymers. As another example of the effect of helical secondary structure on optical activity, we shall use a case similar to that considered in Chapter 8, a dimer in which there is interaction between transition dipoles (see Figure 10.13). This situation resembles that found in polynucleotides. Considering transitions in the planes of the rings, we recall from Chapter 8 that a single transition in the monomer will be split into two transitions in the dimer— a simple form of exciton splitting. These two modes of oscillation are shown in Figure 10.13. Now the transition moments are not coplanar, since the rings,

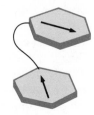

Figure 10.13

A schematic picture of a dimer in which electronic transitions in the two rings are coupled. The rings are parallel but rotated with respect to one another. In-phase and out-of-phase oscillations are shown.

while parallel, have been rotated with respect to one another (the beginning of helical structure). Thus in either mode of oscillation there will be a rotation of charge as well as a translation of charge.† This means that there will be a magnetic dipole as well as an electric dipole associated with each mode. But note also that while in one mode the electric and magnetic moments are in the same direction, in the other they are opposed. This will result in opposite rotational strengths in the two modes, with $R_1 = -R_2$. Therefore, the circular dichroic spectrum will resemble Figure 10.14.

For larger helices, involving larger numbers of residues, each rotated with respect to its neighbor, the splitting of the transitions becomes more complex.

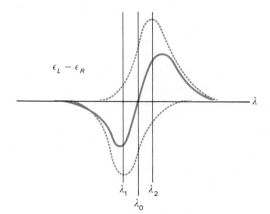

Figure 10.14

The kind of CD spectrum expected for a dimer exhibiting exciton splitting of a transition at λ_0 into two bands at λ_1 and λ_2. The individual CD bands are shown by the broken lines and the overall CD curve by the solid line.

† It will be worthwhile at this point to examine the directions of electron displacement and charge rotation to see how the two modes of oscillation differ.

However, the shape of the CD and ORD curves remains very much the same. In Figure 10.15 is shown the CD spectrum of polyriboadenylic acid, compared to that for the dimer and monomer. Note that the optical activity exhibited by the polymer is much greater and qualitatively different from that of the monomer.

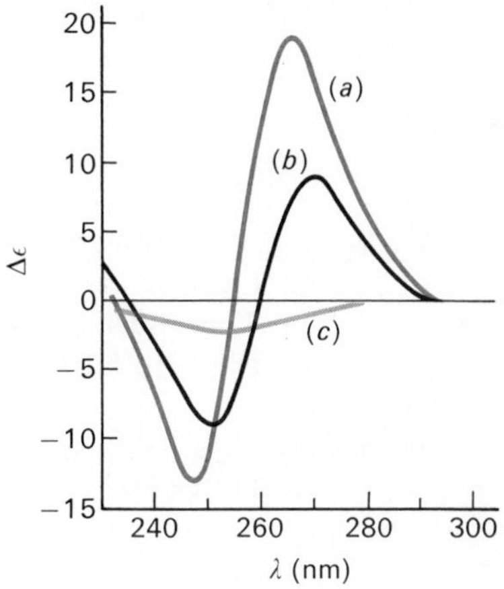

Figure 10.15

Circular dichroism of polyriboadenylic acid. Curve (*a*), high polymer; curve (*b*), dimer; curve (*c*), monomer. All are at neutral pH, with $\Delta\varepsilon$ on a per-mole-of-residues basis. Data taken from K. E. Van Holde, J. Brahms, and A. M. Michelson, *J. Mol. Biol.* **12**, 726 (1965).

It can be shown that if the exciton transitions do not interact with other bands, then rotational behavior will be conservative; that is, we shall expect for the total group of components,

$$\sum_k R_k = 0 \tag{10.8}$$

This situation, in which we have closely spaced series of bands of equal and opposite rotational strength, is approximated in many helical polymers. It is exhibited by the α-helical polypeptides, as shown in Figure 10.16. The backbone transition that in the random coil lies at about 200 nm is split into two series of bands, one at about 207 nm and the other near 190 nm (see also Chapter 8). As can be seen from the CD spectrum in Figure 10.16, these series have opposite rotational strengths. Individual exciton levels are not resolved, so one observes two fairly broad CD bands in the appropriate regions. The situation is complicated by the existence of a weak, but strongly optically active band (probably an $n - \pi^*$ transition)† at about 222 nm.

---

† The $n \longrightarrow \pi^*$ transitions are frequently characterized by weak absorption but high rotational strength. This is because, in taking electrons from the nonbonding orbitals into the π^* orbitals, which lie perpendicular to these, there is little charge displacement (hence small **μ**) but a considerable rotation (large **m**).

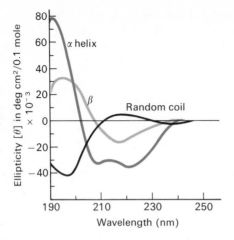

Figure 10.16

Circular dichroism of a polypeptide (poly-l-lysine) in various conformations. From N. Greenfield and G. D. Fasman, *Biochemistry*, **8**, 4108 (1969). Copyright © 1969 by the American Chemical Society.

In contrast to this behavior, the random-coil form exhibits only a single strong dichroic band at about 200 nm., and a very weak positive band near 220 nm. Thus the intensities of the CD bands at 207, 190, or even 222 nm may be used to estimate the percent helix. Table 10.1 summarizes the values found for poly-

TABLE 10.1   OPTICAL CHARACTERISTICS OF REGULAR POLYPEPTIDE STRUCTURES

| Structure | CD Extrema | | ORD Extrema | | Moffitt Parameters | |
	λ (nm)	$[\theta'] \times 10^{-3}$ (deg-cm^2/decimole)	λ (nm)	$[m']$ (deg-cm^2)/decimole	a_0[a]	b_0
Random coil	237	−0.2	205	−15,000	∼ −600	0
	217	+5	190	+17,000		
	197	−42				
α helix	222	−36	233	−15,000	+650	−630
	208	−33	198	+70,000		
	191	+77				
Antiparallel β	217	−18	230	−6,000	400–700	0
	195	+32	205	+26,000		

[a] Depends very much on solvent and amino acid composition. All values are approximate and subject to change.

peptides in various conformations. Note that the β structure (see Barker, in this series) exhibits a still different circular dichroic behavior. Some progress has been made in utilizing these various measures of helicity in the study of globular proteins in solution; Table 10.2 shows some tentative results. It should be emphasized that the actual behavior is complex and not fully understood, so that present estimates must be considered with reservations.

There is yet another way of estimating helix content, which depends on the peculiar form induced in the ORD curve by a complex of transitions obeying the conservative law. We recall from Equation (10.5) that the mean residue

TABLE 10.2   HELICAL CONTENT OF VARIOUS PROTEINS AS CALCULATED BY VARIOUS METHODS

	% Helix (to nearest 5%) from:				
	b_0	a_0	$[m']_{233}$	$[\theta']_{222}$	X ray[a]
Tropomyosin	90	90			
Myoglobin	70–80		60	60–70	70
Bovine serum albumin	45	60		50	
Insulin	40	60		30	
Lysozyme	30	40		35	35
Ribonuclease	15	15		Small	15

[a] From X-ray diffraction studies.

rotation produced by a series of optically active transitions may be written, at wavelengths far from the transitions, as

$$[m'] = A \sum_k R_k \lambda_k^2 / (\lambda^2 - \lambda_k^2) \tag{10.9}$$

Now suppose this is a group of closely spaced transitions, obeying the conservation rule $\sum_k R_k = 0$. At a given wavelength of observation, λ, we can consider (10.9) as the sum of a series of functions of the λ_k's:

$$[m'] = A \sum_k R_k f(\lambda_k)$$

where

$$f(\lambda_k) = \lambda_k^2 / (\lambda^2 - \lambda_k^2) \tag{10.10}$$

Now, all of the λ_k's lie very close to the band center, λ_0. We can then expand the function $f(\lambda_k)$ in a Taylor's series about λ_0. That is,

$$f(\lambda_k) = f(\lambda_0) + (\lambda_k - \lambda_0) f'(\lambda_0) + \cdots \tag{10.11}$$

where $f'(\lambda_0)$ is the first derivative of $f(\lambda_k)$, evaluated at λ_0. Calculating the derivative and inserting into Equation (10.11) and then into (10.9), we obtain

$$[m'] = \frac{A\lambda_0^2}{\lambda^2 - \lambda_0^2} \sum_k R_k + \frac{2\lambda_0 A}{\lambda^2 - \lambda_0^2} \sum_k R_k(\lambda_k - \lambda_0)$$

$$+ \frac{2A\lambda_0^3}{(\lambda^2 - \lambda_0^2)^2} \sum_k R_k(\lambda_k - \lambda_0) + \cdots \tag{10.12}$$

By the conservation rule, the first term vanishes. But the sum $\sum_k R_k(\lambda_k - \lambda_0)$ will not, in general, be zero, so we may write (10.12) in the general form

$$[m'] = \frac{a_0 \lambda_0^2}{\lambda^2 - \lambda_0^2} + \frac{b_0 \lambda_0^4}{(\lambda^2 - \lambda_0^2)^2} + \cdots \tag{10.13}$$

where a_0 and b_0 are constants. This is the famous Moffitt equation. Its importance lies in the fact that it predicts a unique kind of functional dependence [involving $(\lambda^2 - \lambda_0^2)^{-2}$] for the rotatory dispersion of helical polymers. In practice, it has been used empirically, since the optical activity of side-chain residues will contribute to the first, or *Drude*, term. If a value for λ_0 can be guessed, a graph of $[m'](\lambda^2 - \lambda_0^2)$ versus $(\lambda^2 - \lambda_0^2)^{-1}$ will give a straight line, with intercept $a_0 \lambda_0^2$ and slope $b_0 \lambda_0^4$. Such graphs are shown in Figure 10.17;

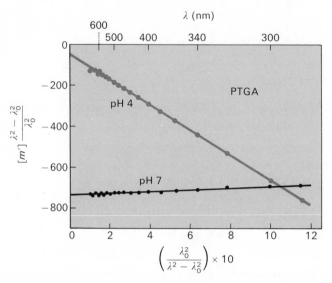

Figure 10.17   Moffitt plots for a polypeptide in α-helical and random-coil conformations. The polypeptide is a copolymer of 5 percent *l*-tyrosine and 95 percent glutamic acid. At pH 4, where the latter residues are mainly uncharged, the polymer is an α helix; at pH 7 the charge repulsion forces it into the random-coil form. From P. Urness and P. Doty (1961).

for the random-coil form, b_0 vanishes, as expected, whereas in the (presumably) wholly helical form, $b_0 = -630$ deg. The empirical a_0 also depends on helix content, for it contains both a contribution from the side-chain residue rotation and the first term of (10.13). Strictly speaking, λ_0 should not be the same for both, but it has become standard practice to use an *average* λ_0 of 212 nm, which appears to work well. Such experiments with polypeptides provide a calibration scale for b_0, which can be then applied to estimate the helix content of the proteins. Some results of such measurements are included in Table 10.2. The Moffitt equation is clearly the most indirect of the techniques we have described for the estimation of helix content. It has the advantage, however, that it utilizes data from the visible and near ultraviolet, where measurements are much easier to make. For this reason, it was widely used in the period before spectropolarimeters which operate at low λ were available. As better equipment became available, more use was made of characteristic values of $[m']$ at extrema, for example, the depth of the minimum at 233 nm (see Table 10.1).

1. *Sketch* the ORD curves that you would expect to correspond to the CD spectra shown below:

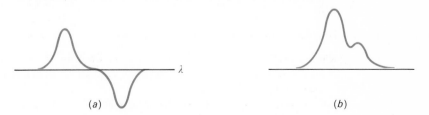

(a)　　　　　　　　　　　　　　　　(b)

*2. Show from the Kronig-Kramer equation that for a very sharp CD band, centered on λ_0, the ORD at a wavelength $\lambda \gg \lambda_0$ is given quite accurately by the Drude equation.

3. The following data describe the circular dichroism of a globular protein. Assuming that the organized structure is exclusively α helix, estimate the percent helix from data in Table 10.1. Does the shape of the CD curve support this assumption?

λ	$[\theta] \times 10^{-3}$ (deg-cm^2/decimole)
240	−1
235	−3
230	−11
225	−18
220	−18
215	−16
210	−20
205	−17
200	−4
195	+25

4. The optical rotatory dispersion of the (deoxygenated) hemocyanin from squid blood is given by the data below:

λ (nm)	$-\alpha$ (deg)
620	0.15
580	0.19
540	0.23
500	0.28
460	0.35
420	0.45
400	0.53
380	0.63
360	0.77
340	0.97

Conditions: 1-cm cell, concentration of 70 mg/ml, and average residue weight = 115. Show that the data fit either a one-term Drude equation (determine λ_0) or a Moffitt equation with $\lambda_0 = 212$ nm. In the latter case, estimate the percent helix. (*Note:* Values of refractive index of water at different wavelengths may be found in the International Critical Tables. Assume that the refractive index for the solutions is the same.)

REFERENCES

Kauzmann, W.: *Quantum Chemistry*, Academic Press, Inc, New York, 1957, Chap. 15 . A lucid discussion of the classic theory of scattering, refraction, and optical rotatory dispersion.

———, *Ibid.*, Chap. 16. Much the same material but from the quantum-mechanical viewpoint.

Schellman, J., and C. A. Schellman: in *The Proteins* (H. Neurath, ed.), Vol. II, Academic Press, Inc., New York, 1964, p. 1. A good, though brief description of the relationship between polymer structure and optical rotary power.

Urness, P., and P. Doty: *Advan. Protein Chem.*, **16**, 401 (1961). Somewhat dated now but excellent as a survey of early ORD work.

ELEVEN | X-RAY DIFFRACTION

One of the most common, if unnoticed, examples of diffraction is the pattern of a cross formed by a distant street light seen through a window screen or through a stretched pocket handkerchief. If the screen is sufficiently fine and if the pattern is examined closely, it will be seen to be made up of a rectangular array of spots of light of varying intensities. This is simply the two-dimensional diffraction pattern of the light by the two-dimensional grid of the screen. Each spot can be identified by counting to the right or left and up or down from the center of the pattern and by assigning these two integers as the "indices" (h, k) of the spot.

With a finer mesh screen the pattern will be seen to be expanded, and with a coarser mesh the spots will be closer together. This reciprocal relationship between diffracting object and diffraction pattern is universally valid, and the coordinate system in which the pattern is measured is referred to as "reciprocal space." If, instead, one now chooses a screen of the same mesh but with a distinctive pattern such as would be produced by adding a finer wire alongside each coarse one to give a Scotch plaid effect, one finds that the spots in the diffraction pattern are in the same places but that their relative brightness has been changed. These experiments demonstrate two fundamental facts about diffraction: The size of the mesh determines the positions of the spots but the nature of the weave determines their intensities.... (R. E. Dickerson, 1964).

In this chapter, we shall be concerned with the kind of information that can be obtained from examination of the "mesh and weave" of crystals of macromolecules.

Probably no other physical technique has had more impact, actual or potential, on biochemistry than X-ray diffraction. The results of these applications in one area alone, the study of globular proteins, provides a startling example. In the space of 10 years, we have progressed from a situation in which these molecules were only vaguely visualized, and experimental studies were literally probing in the dark, to a position in which we *know*, for a number of proteins, the precise spatial location of all of the residues in the chain. Now, and only now, studies of the active sites of enzymes become meaningful, the significance of evolutionary changes can be discussed intelligently, and the interactions of protein molecules with other macromolecular structures can be visualized.

In discussing X-ray diffraction, we shall adopt an approach a little different from that which we have used in earlier chapters. X-ray diffraction is a complex and difficult technique, and it is unlikely that the average reader will employ it himself, as he might sedimentation or light scattering. Even to provide a comprehensive picture of the theory would require more space than we can allow. We shall be content to define the general problems, indicate the limitations and difficulties in the experimental technique, and illustrate the results by a few examples. A fuller description of the theory and methods may be found in some of the references, and the results are cited again and again in other volumes of this series.

11.1 FUNDAMENTALS OF X-RAY DIFFRACTION

Most elementary discussions of X-ray diffraction begin with a "derivation" of the Bragg law. A crystal is pictured as a set of reflecting planes (Figure 11.1) with uniform spacing from which X rays, incident at an angle θ, are reflected at an angle θ. It is then argued that for reinforcement of the reflected waves to

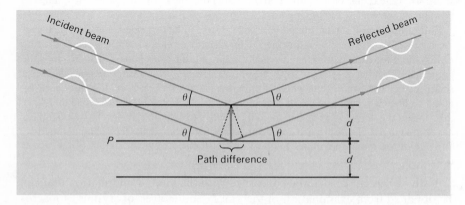

Figure 11.1 A conventional derivation of the Bragg equation. The crystal planes P are considered to be reflectors. Reinforcement in the reflected beam will occur only when the path difference is some multiple of the wavelength, λ, so θ must satisfy the equation $n\lambda = 2d \sin \theta$.

occur, the angle θ must satisfy the condition

$$2d \sin \theta = n\lambda \tag{11.1}$$

where n is an integer. If the crystal is large and perfect, containing many parallel planes, destructive interference will result at all angles not stringently satisfying this condition.

Even a glance at such a derivation should provoke skepticism in an alert student. What in the world is doing the reflecting? Are not the "reflecting planes" really made up of layers of atoms or molecules, with more empty space than matter? These objections are quite valid, and yet the Bragg equation is correct. Let us see how a rational derivation can be arrived at.

In the first place, X-ray diffraction is a scattering phenomenon. Its unique characteristics arise because (1) the scattering is from a periodic structure, and (2) the wavelength (λ) of the X rays is comparable to the periodic spacing of the structure. In Chapter 9 we saw that the scattering from a periodic structure with spacings much smaller than the wavelength was largely eliminated by internal interference. If the wavelength and spacing are comparable, interference and reinforcement are both possible. Let us consider X-ray scattering from the simplest periodic structure, a one-dimensional row of scattering centers (atoms, for example) with uniform spacing (Figure 11.2). The direction

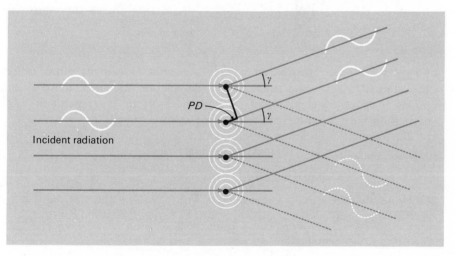

Figure 11.2   The diffraction from a row of scattering centers. Reinforcement will be found at an angle where PD = $l\lambda$ (l is an integer).

of incident radiation is taken to be perpendicular to the row of atoms. If one now imagines scattering from the centers, this scattering will be in all directions, as is light scattering. However, only in certain directions will reinforcement of scattered rays occur; those will be such that the path difference (PD) between rays scattered by adjacent atoms corresponds to an integral number of wavelengths. If γ is the scattering angle, and c the spacing:

$$l\lambda = c \cos \gamma \qquad (11.2)$$

where l is an integer. If the row of atoms is very long, complete annihilation will occur at all other angles. Then, from such a row, radiation will be scattered only along conical surfaces, as shown in Figure 11.3. A screen or photographic plate placed perpendicular to the incident radiation will intersect these cones

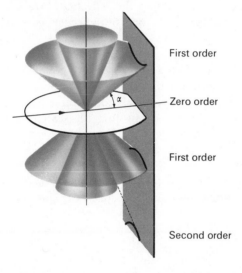

First order

Zero order

First order

Second order

Figure 11.3
The cones along which constructive interference can occur for a row of scatterers. These are shown intersecting a photographic plate. On the plate, layer lines corresponding to $l = 0$, 1, 2, and so forth will appear.

on hyperbolic lines. The spacing of these *layer lines* on the plate will be nearly proportional to $\cos \gamma$, and hence to $1/c$. This is our first encounter with the reciprocal space concept; closer spacings in a crystal correspond to large spacings on the diffraction pattern and vice versa.

If the incident radiation makes an angle γ_0 other than 90 deg with the row of scatterers, Equation (11.2) must be modified† to

$$l\lambda = c(\cos \gamma - \cos \gamma_0) \qquad (11.3)$$

If a two-dimensional array is considered, with spacing a and c in the x and z directions, respectively, two equations must be satisfied simultaneously:

$$l\lambda = c(\cos \gamma - \cos \gamma_0) \qquad (11.4)$$

$$h\lambda = a(\cos \alpha - \cos \alpha_0) \qquad (11.5)$$

Physically, this means that reinforcement will now occur only along the lines where the cones of diffraction from the x rows intersect those from the z rows. See Figure 11.4. Spots will appear on a photographic plate placed parallel

---

† Proof of this is not difficult. It is left to the reader, to promote a better understanding of the equations.

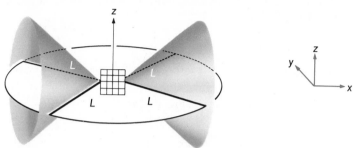

Figure 11.4   Diffraction by a rectangular lattice in the xz plane. For clarity, only one set of cones of diffraction produced by the x spacings is shown, and only the circle (zero-order reflection) from the z axis is shown. Diffraction will be possible along the lines (L) where these surfaces intersect. These could be called the 1, 0 reflections.

to the two-dimensional array. This pattern is the kind referred to in the quotation from Dickerson.

If a three-dimensional array, such as an orthorhombic crystal, with spacings a, b, and c in the x, y, and z directions is employed, a third equation must also be satisfied:

$$k\lambda = b(\cos \beta - \cos \beta_0) \tag{11.6}$$

These three equations [(11.4) through (11.6)], are called the von Laue equations.

For arbitrary orientations of the crystal, it is not possible to satisfy the von Laue equations simultaneously.† Only when the crystal takes certain orientations with respect to the incident beam can reinforcement occur. Suppose that the crystal is arbitrarily oriented with respect to the beam, and the following question is asked: If we consider an angle 2θ with respect to the incident beam, is there reinforcement in this direction? If so, the von Laue equations must be satisfied. To simplify the nomenclature, we write $\underline{\alpha}$ for $\cos \alpha$, $\underline{\beta_0}$ for $\cos \beta_0$, and so forth. Note that these are also the *direction* cosines for the diffracted and incident beam with respect to the crystal axes. For direction cosines of a line the following relations apply:

$$\underline{\alpha}^2 + \underline{\beta}^2 + \underline{\gamma}^2 = 1 \tag{11.7}$$

$$\underline{\alpha_0}^2 + \underline{\beta_0}^2 + \underline{\gamma_0}^2 = 1 \tag{11.8}$$

and if the one line is making an angle 2θ with respect to the projection of the other, we have

$$\cos 2\theta = \underline{\alpha\alpha_0} + \underline{\beta\beta_0} + \underline{\gamma\gamma_0} \tag{11.9}$$

† This is not difficult to see geometrically. While two nonparallel surfaces will always intersect in a line or lines, three surfaces will mutually intersect along a single line only in special circumstances.

These are simply standard geometric relationships. If we square each of the von Laue equations, we find

$$\frac{h^2\lambda^2}{a^2} = \underline{\alpha}^2 - 2\underline{\alpha}\alpha_0 + \underline{\alpha}_0^2 \tag{11.10}$$

$$\frac{k^2\lambda^2}{b^2} = \underline{\beta}^2 - 2\underline{\beta}\beta_0 + \underline{\beta}_0^2 \tag{11.11}$$

$$\frac{l^2\lambda^2}{c^2} = \underline{\gamma}^2 - 2\underline{\gamma}\gamma_0 + \underline{\gamma}_0^2 \tag{11.12}$$

or, summing,

$$\left(\frac{h^2}{a^2} + \frac{k^2}{b^2} + \frac{l^2}{c^2}\right)\lambda^2 = 1 - 2(\underline{\alpha}\alpha_0 + \underline{\beta}\beta_0 + \underline{\gamma}\gamma_0) + 1$$

$$= 2(1 - \cos 2\theta) \tag{11.13}$$

By a standard identity, $1 - \cos 2\theta = 2 \sin^2 \theta$; therefore,

$$\lambda^2\left(\frac{h^2}{a^2} + \frac{k^2}{b^2} + \frac{l^2}{c^2}\right) = 4 \sin^2 \theta$$

or

$$\lambda\left(\frac{h^2}{a^2} + \frac{k^2}{b^2} + \frac{l^2}{c^2}\right)^{1/2} = 2 \sin \theta \tag{11.14}$$

This is simply a generalization of the Bragg law [Equation (11.1)] to an orthorhombic crystal with spacings a, b, and c.†

Thus the Bragg law is proved, starting with the concept of diffraction as a scattering phenomena. We have *not* shown, however, that the angle θ is in fact a reflection angle from some set of crystal planes. Such is demonstrated, for example, in Lipson and Cochran, 1953.

The fact that arbitrary orientations do not allow satisfaction of the von Laue equations requires that we examine a crystal over a range of orientations to observe its diffraction pattern. One common way of doing this is called the *rotating crystal* method. In Figure 11.5 are shown a set of unit cells with indices for some of the sets of crystal planes shown. These are the Miller indices: The 010 planes are perpendicular to the y axis. If the crystal is oriented as shown in Figure 11.6 and then rotated about the z axis, diffraction from the

---

† For example, for 100 planes, we obtain $h\lambda/a = 2 \sin \theta$. The expression in brackets in Equation (11.14) will, in general, represent the reciprocal of the distance between planes with Miller indices h, k, l. If the crystal is not orthorhombic, the expression corresponding to (11.14) is more complicated.

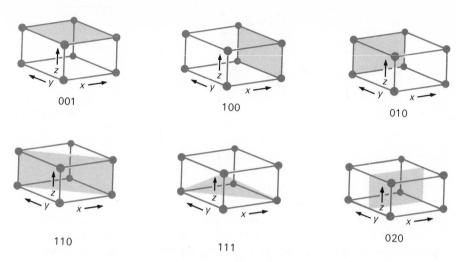

Figure 11.5   Miller indices of some of the possible planes in an orthorhombic unit cell (a rectangular parallelepiped) with sides a, b, c. The numbering rule is as follows: If the plane cuts an axis at a/h, b/k, c/l, the indices are h, k, l.

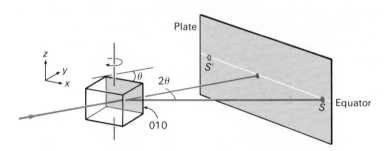

Figure 11.6   The rotating crystal method. The crystal has been oriented with the z axis perpendicular to the beam and is rotated about this axis. When the 010 planes make the Bragg angle θ with respect to the beam, a diffraction spot(S) is produced on the equator of the image at an angle 2θ from the incident beam. Another spot will be observed at S', when the angle is $-\theta$. Higher-order reflections will also be seen. Higher-order spots will be farther out on the pattern.

010 planes will be possible only when these planes make an angle θ with the direction of the beam; this angle must be one that satisfies the Bragg equation for this set of planes.† A corresponding spot will appear on the photographic plate. Other spots will appear on this layer line for higher orders of diffraction ($h = 2, 3$, and so forth) and for higher-order spacings (020 planes and the like). Note that in keeping with the reciprocal space concept, those spots corresponding to larger spacings will appear closer to the center of the pattern. The

---

† At this point it would be very useful for the reader to convince himself that the individual von Laue equations are all satisfied in the orientation shown in Figure 11.6.

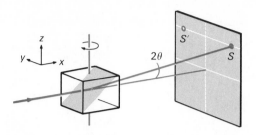

Figure 11.7

Like Figure 11.6 but showing how off-equator spots are given by planes not parallel to the z axis.

generation of spots on other layer lines is depicted by Figure 11.7. Here, the 101 planes are so oriented as to satisfy the Bragg condition. In this way, rotation of the crystal in the beam will serve to produce an image that contains many of the diffraction maxima possible for the structure.†

From the spacing of spots in the pattern, the *unit cell* dimensions can be deduced. The unit cell is that smallest portion of the crystal that, by continued repetition, yields the entire structure. In the simple orthorhombic crystal we considered initially, it is a rectangular parallelepiped with edges *a, b, c*.

There is obviously another way in which a sample can be assured to satisfy the Bragg conditions. If, instead of a single crystal, we utilize a powder made up of many small, randomly oriented crystals, some will surely be so oriented as to satisfy the Bragg condition for each set of crystal planes. But the crystals will also be randomly oriented with respect to the axis defined by the incident beam. This will smear each spot on the diffraction pattern into a circle about the origin (imagine rotating the crystal in Figure 11.6 about the beam axis). Thus such a *powder pattern*, while giving the lattice spacings, does not allow us to identify them in the way that they can be from a single-crystal photograph. See Figure 11.8.

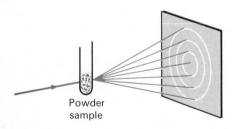

Powder
sample

Figure 11.8
The powder method. See the text.

11.2 DIFFRACTION BY FIBROUS MACROMOLECULES

A situation somewhat between that of a single crystal and a powder is found in the X-ray diffraction of fibers. A fiber may possess some regularity of structure along the fiber axis, with more or less regularity in directions perpendicular to the axis. Since individual crystallites in the fiber will be more or less parallel

† It should be obvious that the crystal must be rotated about more than one axis to yield all of the reflections.

to the fiber axis, but with various orientations *about* this axis, the same situation is achieved as by rotation of a single crystal. Depending on the extent of orientation in the fiber,† the diffraction pattern will exhibit rings, arcs, or spots (see Figure 11.9). The diffraction pattern exhibited by the β structure of silk fibroin

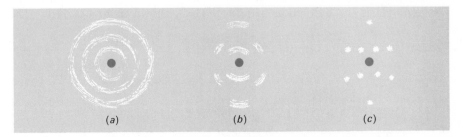

(a)   (b)   (c)

Figure 11.9   Diffraction from a fiber in which molecules are (a) unoriented, (b) partly oriented, and (c) highly oriented.

is shown in Figure 11.10. Note that on the meridian (the vertical line corresponding to the fiber axis) a strong reflection is seen at 3.5 Å, corresponding to

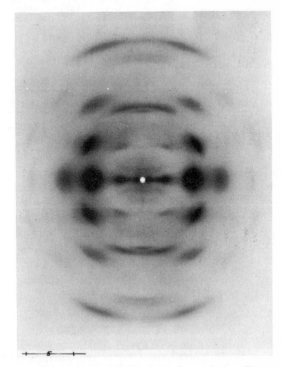

Figure 11.10

Fiber diffraction pattern of silk fibroin. From R. E. Dickerson, 1964.

† Some biopolymers will produce oriented fibers when a thread is drawn from solution. Sometimes orientation can be further increased by stretching.

the amino acid repeat distance. Similarly, on a level corresponding to a two-residue repeat (7 Å) are seen two spots. These lie off the meridian because of the "pleating" of the sheet. The equatorial reflections (see Figure 11.6) correspond to spacings between the chains in different directions.

If a fibrous material is helical, a very typical type of diffraction pattern is found, characterized by diagonal rows of spots proceeding from the origin. An example is the diffraction pattern of DNA shown in Figure 11.11. The

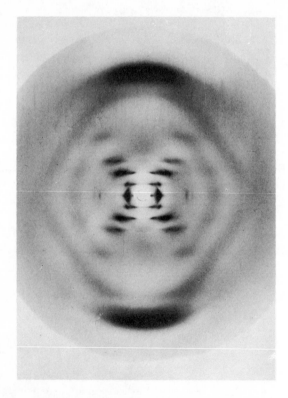

Figure 11.11

X-ray diffraction pattern from wet DNA fiber. From R. E. Dickerson, 1964.

reciprocals of the vertical distances of spots from the equatorial plane correspond, of course, to repeat distances along the helix axis. The angle made by the "X" pattern with the equatorial plane very nearly equals the pitch angle of the helix. A more refined analysis, of course, gives much more detail about the helical structures.

Studies of fibrous proteins, polypeptides, and polynucleotides have provided evidence for a remarkable variety of such structures. For further details, the reader is referred to other volumes in this series or to the references given at the end of this chapter.

11.3   DIFFRACTION BY MOLECULAR CRYSTALS; THE PHASE PROBLEM

Analysis of the scattering by crystals composed of complex molecules such as globular proteins is a much more difficult task than the determination of periodicities and spacings in fibrous structures. The molecule of an enzyme such as cytochrome (Figure 11.17) does not involve a simple, repetitive, three-dimensional structure, for the chain is wound and folded in a complex way.

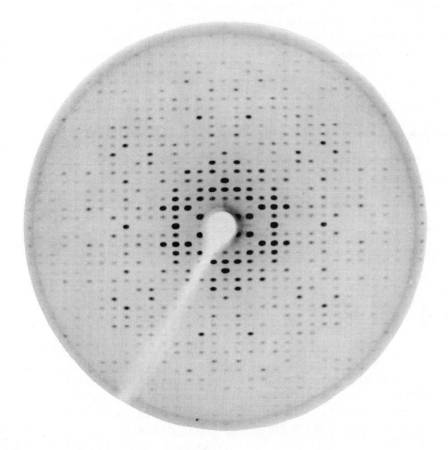

Figure 11.12   X-ray diffraction pattern from a crystal of horse heart oxidized cytochrome *c*.
Courtesy of R. E. Dickerson.

The enormous number of reflections observed in the diffraction pattern attests to the regularity of conformation of molecules in the crystal lattice.† To deduce the detailed structure of such a molecule, it is clearly necessary to use as much as possible of the information available from the diffraction photograph. This means not only the spacings but also the intensities of the spots. So we must ask, what determines the relative intensities of the spots?

---

† The regularity is sometimes astonishing. Good data on myoglobin have been attained to a resolution of 1.4 Å. This means that the great majority of the atoms in a great majority of the molecules in the crystals are held with at least this precision.

Consider a unit cell, which we shall assume contains one protein molecule. This is an arrangement of atoms (scattering centers) that are separated by distances comparable to the wavelength of the X rays. We can define a set of crystallographic planes by picking the corresponding atom in each molecule and passing the planes through them. Since there is only one molecule per cell, these planes define the unit cell. Any other set of corresponding atoms lie on a similar but displaced set of crystallographic planes. There will be interference between the scattering between these sets of planes. That is, while all sets will give exactly the same set of reflections, their contributions to these reflections will differ in *phase*.

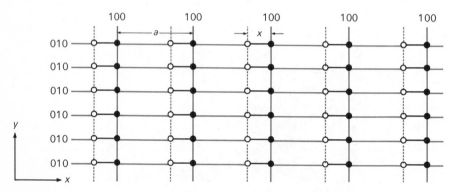

Figure 11.13   The *xy* plane of a hypothetical lattice of diatomic molecules. Note that the same lattice (displaced) may be used either for the white atoms or the black atoms. The molecules are assumed to lie in the plane of the paper. See the text.

To make this argument more concrete, consider Figure 11.13, which depicts the simplest possible molecular crystal—one made up of diatomic molecules. Let us assume that all the molecules lie in planes parallel to the *xy* plane and oriented as shown in the figure. Suppose that the crystal is oriented in the X-ray beam so that the Bragg condition [Equation (11.1)] is satisfied for the 100 planes. We draw a set of such planes (perpendicular to the paper) through the *black* atoms. Note that if the Bragg (and hence von Laue) equations are satisfied, these atoms are scattering in phase in the direction of the reflected beam. This means the phase difference between planes of black atoms is some multiple of 2π rad (360 deg). Note, also, that the *white* atoms are doing precisely the same thing and are in phase with one another. *But the scattering from the white atoms is out of phase with that from the black.* How much? If the *spacing* in the *x* direction is *a*, and the displacement of the white lattice from the black lattice is *x*, the phase difference must be

$$\Delta\phi = 2\pi\frac{x}{a} \tag{11.15}$$

for the first-order reflections. For higher orders, we have $2\pi hx/a$, where $h = 1, 2, 3, \ldots$. For reflection from any arbitrary set of planes, with indices h, k, l, we have

$$\Delta\phi = 2\pi\left(\frac{hx}{a} + \frac{ky}{b} + \frac{lz}{c}\right) \tag{11.16}$$

where one atom has been chosen to be at the origin of the unit cell, $x = 0, y = 0$, $z = 0$. In Chapter 9 we considered the problem of calculating the amplitude and intensity of the radiation scattered from a pair of oscillators. Here we are dealing with a generalization of that problem, the scattering from a group of atoms that make up the contents of the unit cell. While the calculation could be done in terms of sines and cosines, as in Chapter 9, this becomes very un-wieldy if many waves of different amplitudes (from centers of differing scattering power) are to be added. The problem becomes much simpler when waves are written in terms of complex numbers. We may represent a periodic displace-ment of amplitude f_k and phase $\Delta\phi_k$ either by

$$f_k \cos(2\pi vt + \Delta\phi_k) \quad \text{or} \quad f_k e^{i(2\pi vt + \Delta\phi_k)}$$

where $i = \sqrt{-1}$. The representations are physically equivalent, for since $e^{i\theta} = \cos\theta + i\sin\theta$, the real part of the expression to the right (which is the physically meaningful part) is identical to the expression on the left. Thus, if we have a jumble of waves at the point of observation, the total field will be

$$E = \sum_k f_k e^{i(2\pi vt + \Delta\phi_k)}$$

$$= \sum_k f_k e^{i2\pi vt} e^{i\Delta\phi_k} = e^{i2\pi vt} \sum f_k e^{i\Delta\phi_k} \tag{11.17}$$

The amplitude of the sum of a number of waves of different phase and amplitude is seen, from (11.17), to be

$$F = \sum f_k e^{i\Delta\phi_k} \tag{11.18}$$

If the scattering centers are atoms, the f_k's will be proportional to the scattering powers of these atoms. If the scattering centers are taken to be small volume elements in the unit cell, the f_k's are proportional to the electron densities in these elements. Returning now to the calculation of the scattering from a lattice of black and white atoms, we may write immediately

$$F_{100} = f_B e^{i\Delta\phi_B} + f_W e^{i\Delta\phi_W} \tag{11.19}$$

where the subscripts B and W refer to black and white atoms, respectively.

Since we have taken the black lattice as centered, we have

$$F_{100} = f_B e^{2\pi i(0/a)} + f_W e^{2\pi i(x/a)}$$

$$= f_B + f_W e^{2\pi i x/a} \tag{11.20}$$

Now, the intensity of the reflection will be proportional to the square of the amplitude (or to the product of the amplitude and its complex conjugate,† if F is complex):

$$I(hkl) = F(hkl)F^*(hkl) \tag{11.21}$$

or

$$I(hkl) = F^2(hkl) \quad \text{if } F \text{ is real.}$$

In our example,

$$I(100) = f_B^2 + f_B f_W (e^{2\pi i x/a} + e^{-2\pi i x/a}) + f_W^2 \tag{11.22}$$

If we use the identity $e^{\pm i\theta} = \cos\theta \pm i\sin\theta$, we can express the middle term as a cosine:

$$I(100) = f_B^2 + 2f_B f_W \cos\frac{2\pi x}{a} + f_W^2 \tag{11.23}$$

If $x/a = 0$ or 1 (atoms superimposed), we have $I(100) = (f_B + f_W)^2$. Otherwise the result is smaller; there is destructive interference. In fact, in the special case where $x/a = \frac{1}{2}$, we obtain $\cos 2\pi x/a = -1$, and $I(100) = (f_B - f_W)^2$. If $f_B = f_W$ (the atoms have the same scattering power), this would mean that the intensity of the 100 reflection would be zero. This special case can easily be seen another way; the atoms would simply cancel out one another's scattering in this reflection. The result is thus seen to be a general case of the example in Chapter 9.

This lengthy discussion of a simple problem points out the complexities facing the protein crystallographer. Each reflection can involve thousands of contributions from the atoms in the unit cell, each differing in phase. Equation (11.18) can be generalized to

$$F(hkl) = \iiint\limits_{\text{unit cell}} \rho(x, y, z)e^{2\pi i[(hx/a) + (ky/b) + (lz/c)]} \, dV \tag{11.24}$$

where $\rho(x, y, z)$ is the electron density in the volume element dV in the unit cell. The situation would be hopeless but for one fact. The crystal is a periodic

---

† The conjugate of a complex number is the number with i replaced by $-i$. Thus in (11.20) $F_{100}^* = f_B + f_W e^{-2\pi i x/a}$.

structure; therefore $\rho(x, y, z)$ is a periodic function—it must repeat from cell to cell. Equation (11.24) will be recognized by any student acquainted with Fourier series. It is simply the general expression for the coefficients in a Fourier series representation of the periodic function $\rho(x, y, z)$. Thus the electron density may be written as

$$\rho(x, y, z) = \frac{1}{V} \sum_{h=-\infty}^{\infty} \sum_{k=-\infty}^{\infty} \sum_{l=-\infty}^{\infty} F(hkl) e^{-2\pi i[(hx/a) + (ky/b) + (lz/c)]} \tag{11.25}$$

where V is the volume of the unit cell. Thus, if we knew the $F(hkl)$'s (which are called the *structure factors*), we would be able to calculate $\rho(x, y, z)$ directly and obtain a picture of the unit cell contents (the molecule!). But there is a "rub." We can only measure intensities, and these give at best $F^2(hkl)$. We know the magnitudes but not the phases. That is, we know the magnitude of a complex quantity but not its phase. The problem can be expressed succinctly by recalling another way of representing a complex number—by a vector on the complex plane (Figure 11.14). We know the length of the vector but not its direction.

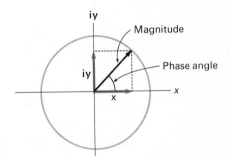

Figure 11.14

A complex number represented as a vector in the complex plane. Note that it is indeed the vector sum of vectors **x** and *i***y** on the axes.

Solutions to this *phase problem* take two general forms. For simple molecules, one can guess at the structure. This gives trial $F(hkl)$'s complete with phase, from which a diffraction pattern can be computed. Comparison of this with the observed pattern may suggest refinements. This procedure is out of the question with protein molecules, but it has been useful both with small biologically important structures such as vitamins and in refining the structures of fibrous biopolymers. A second general approach is known as the *isomorphous replacement* method. This involves adding heavy atoms (strong scatterers) to the protein molecule and investigating their effect on the diffraction. The heavy atoms must not modify either the protein conformation or the crystal form; the replacement must be isomorphous.

Now it is not too difficult to locate one or more heavy atoms in the unit cell (see the Patterson method, below). If a heavy atom is located, then the amplitude and phase of its scattering are known. We then note that both $F(hkl)$, the structure factor for the protein, and $f_H(hkl)$, the structure factor for the heavy atom, can be represented as vectors on a map, as in Figure 11.14. Furthermore, if the heavy atom scatters much—much more than the light atom it replaces—$F'(hkl)$, the corresponding structure factor for the modified protein, must be represented as a vector that is the vector sum of the other two. The problem now is as follows: We have three vectors of known magnitude; one is the sum of two others, and the phase of one of these is known. The problem can be nearly solved geometrically (see Figure 11.15). But there are usually two acceptable solutions. Use of a *second* heavy atom derivative will then allow a choice between these two.

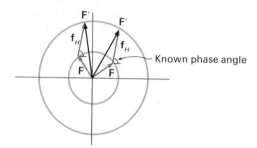

Figure 11.15

The heavy atom method for getting phases. Magnitudes of all vectors and the phase of $\mathbf{f}_H$ are known. There are two possible solutions for $\mathbf{F}$, $\mathbf{F}'$. See the text.

A useful technique for the determination of heavy atom positions involves Patterson maps. A function $P(hkl)$ is written as

$$P(x, y, z) = \frac{1}{V} \sum_{h=-\infty}^{\infty} \sum_{k=-\infty}^{\infty} \sum_{l=-\infty}^{\infty} F(hkl)F^*(hkl)e^{-2\pi i[(hx/a)+(ky/b)+(lz/c)]} \qquad (11.26)$$

Note that this is not too difficult to calculate; it contains only the squares of structure factors, which are given by the intensities of the various spots. The Patterson map turns out to be a measure of densities of *distances* between groups. It is most easily visualized by considering a unit cell containing three scatterers [Figure 11.16(a)]. If we measure all vectors between these centers, move these vectors to the origin, and graph the positions of their heads, we will have the resulting Patterson map [Figure 11.16(b)]. Note that

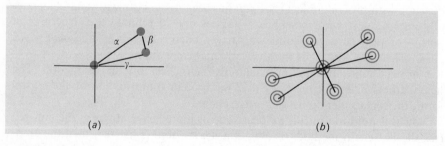

(a)   (b)

Figure 11.16   A planar three-atom "molecule" (a) and its corresponding Patterson function (b). See the text.

it is more complex than the model itself. For proteins, then, the complete Patterson is almost useless. But a "difference Patterson," obtained by comparison of the diffraction patterns of native and heavy-atom-substituted protein, is much simpler and is of great aid in locating the heavy atom in the unit cell.

In brief, then, the procedure for deducing the three-dimensional structure of a protein proceeds somewhat as follows:

1. Good crystals of the protein must be prepared. Preliminary studies yield crystal form and unit cell.

2. Several isomorphous replacements must be obtained.

3. Diffraction patterns of a number of these must be obtained and measured.

4. Heavy atom positions must be deduced, and the phases of reflections must be calculated.

5. A trial electron density map is obtained. This may be used to back-calculate spot intensities and refine the calculation.

(a)   (b)

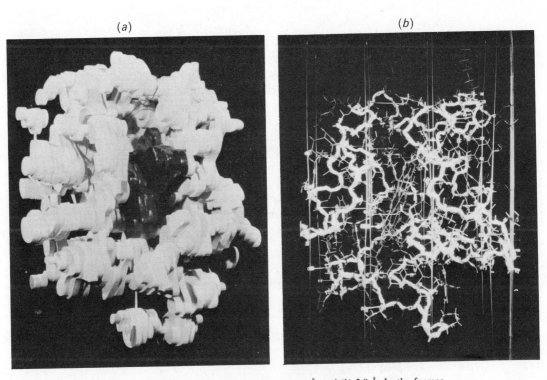

Figure 11.17   Models of cytochrome c at resolution of (a) 4 Å and (b) 2.8 Å. In the former the chain can barely be traced. In the latter it is possible to see the orientation of individual amino acid residues. Courtesy of R. E. Dickerson.

At this point, one aspect of the reciprocal space concept must be reiterated. The reflections closest to the X-ray beam correspond to the largest regular spacings; those farther out yield finer details. It can be put another way: Those spots with smallest indices (*hkl*) yield the first coefficients in the Fourier series (11.25). Thus a first approximation to the structure can be obtained by choosing only a relatively small number of reflections (a few hundred on each isomorphous derivative). Further refinement can then be made by taking into account more reflections. At each point, one can speak of the "resolution" of the crystal structure. The models of cytochrome *c* shown in Figure 11.17 illustrate the effect of enhanced resolution. This process cannot be continued indefinitely; for one thing, it gets expensive, the number of reflections needed going up roughly as the inverse square of the resolution. Furthermore, the diffraction patterns do not yield infinite detail. They fade out for small spacings, perhaps because the protein structure itself exhibits variation below certain limits. However, that this detail may be quite fine is shown by the resolution obtained in Figure 11.17.

11.4 X-RAY DIFFRACTION STUDIES ON GLOBULAR PROTEINS: A SUMMARY AND PROSPECTIVE

The preceding paragraphs have grossly underemphasized the experimental difficulties in this kind of research. Each structure determination is a vast project, often taking years of work by a skilled team. Yet so great is the impact of the results that have been obtained that at this writing about 10 such structures have been elucidated in high resolution. You will read about the interpretations of some of these data in other volumes in this series.

Methods are being constantly improved, and it may be confidently expected that many more structures will be resolved in the next decade. In a sense this achievement represents the culmination of a phase of molecular biology, for now that we know exactly *what* these molecules are like, we may begin to understand *how* they function, both individually and in the complex assemblages that are cells and organisms. We can begin to see how, in D'Arcy Thompson's words, they have been "moved, moulded, and conformed."

REFERENCES

The following is a standard reference for X-ray diffraction analysis:

Lipson, H. and W. Cochran: *The Determination of Crystal Structures*, G. Bell & Sons, Ltd., London, 1953.

On a more elementary level:

Wheatley, P. J.: *The Determination of Molecular Structure*, Oxford University Press, New York, 1959, Chaps. 1 and 6–8.

For applications to biopolymers:

Crick, F. H. C. and J. C. Kendrew: "X-Ray Analysis and Protein Structure," *Advan. Protein Chem.*, **12**, 133 (1957).

Dickerson, R. E.: in *The Proteins* (H. Neurath, ed.), Academic Press, New York, 1964, Chap. 11.

Stryer, L.: "Implications of X-Ray Crystallographic Studies of Protein Structure," *Ann. Rev. Biochem.*, **37**, 25 (1968).

INDEX